MAPPING THE DYNAMICS OF SCIENCE AND TECHNOLOGY

Also by John Law

POWER, ACTION AND BELIEF: A New Sociology of Knowledge? (*editor*)

SCIENCE FOR SOCIAL SCIENTISTS (*with Peter Lodge*)

Also by Arie Rip

MACHT OVER KENNIS (*joint editor*)

WETENSCHAP ALS MENSENWERK

DISTRIBUTED BY
SHERIDAN HOUSE INC.
145 PALISADE STREET
DOBBS FERRY, NY 10522

MAPPING THE DYNAMICS OF SCIENCE AND TECHNOLOGY

Sociology of Science in the Real World

Edited by

Michel Callon

Professor of Sociology, École des Mines de Paris
Director, Centre de Sociologie de l'Innovation

John Law

Senior Lecturer and Head of Department of Sociology, Social Anthropology and Social Work, University of Keele

Arie Rip

Professor of Science Dynamics
University of Amsterdam

M

First published 1986

Published by
THE MACMILLAN PRESS LTD
Houndmills, Basingstoke, Hampshire RG21 2XS
and London
Companies and representatives
throughout the world

Filmset by Vantage Photosetting Co. Ltd.,
Eastleigh and London
Printed in Hong Kong

British Library
Cataloguing in
Publication Data

Mapping the dynamics of science and technology.
1. Science
I. Callon, Michel   II. Law, John, 1946–
III. Rip, Arie
500'.72   Q181
ISBN 0–333–37223–9

Contents

List of Figures vii
List of Tables ix
Notes on the Contributors x
Acknowledgements xiii
Glossary xvi

PART I INTRODUCTION

1 How to Study the Force of Science 3
Michel Callon, John Law and Arie Rip

PART II THE POWER OF TEXTS IN SCIENCE AND TECHNOLOGY

2 The Sociology of an Actor-Network: The Case of the Electric Vehicle 19
Michel Callon

3 Laboratories and Texts 35
John Law

4 Writing Science – Fact and Fiction: The Analysis of the Process of Reality Construction Through the Application of Socio-Semiotic Methods to Scientific Texts 51
Bruno Latour and Françoise Bastide

5 The Heterogeneity of Texts 67
John Law

6 Mobilising Resources Through Texts 84
Arie Rip

PART III MAPPING SCIENCE AND TECHNOLOGY

7 Qualitative Scientometrics 103
Michel Callon, John Law and Arie Rip

8 Aquaculture: A Field by Bureaucratic Fiat 124
Serge Bauin

9 State Intervention in Academic and Industrial Research: The Case of Macromolecular Chemistry in France 142
William Turner and Michel Callon

10 Pinpointing Industrial Invention: An Exploration of Quantitative Methods for the Analysis of Patents 163
Michel Callon

11 Technical Issues and Developments in Methodology 189
Jean-Pierre Courtial

12 Future Developments 211
Michel Callon, Jean-Pierre Courtial and William Turner

PART IV CONCLUSION

13 Putting Texts in Their Place 221
Michel Callon, John Law and Arie Rip

Bibliography 231
Index 239

List of Figures

2.1	The anatomy of the EDF actor-world	29
3.1	Blood clearance of ^{125}I-labelled copolymers 29–32	44
3.2	Body distribution of radioactivity sixty minutes after intravenous injection	45
4.1	Turnover, with time, of water and sodium in the different regions of the kidney	55
5.1	The elementary form of interessement	72
5.2	Linked translations or the funnel of interests	79
7.1	An article as a network of powerful words	106
7.2	The network of problematisations around a single paper	109
7.3	The network of problematisations in an area of science	109
7.4	Visual representation of inclusion index	113
7.5	Hierarchical relationship between problematisations as depicted by inclusion index	113
7.6	Clusters of problematisations as generated by the proximity index	115
7.7	Visual representation of proximity index	115
7.8	Inclusion map for dietary fibre research, 1973–6	118
7.9	Inclusion map for dietary fibre research, 1977–8	119
7.10	Proximity map of aquaculture, 1979	120
7.11	Proximity map of breeding and cultivation cluster	121
8.1	Inclusion map (detail), 1979	126
8.2	Main central and mediator words, 1979	127
8.3	Main central and mediator words, 1981	128
8.4	Detail from 1979 inclusion map (*Salmo gairdneri*)	128
8.5	Detail from 1979 inclusion map (algae)	129
8.6	Detail from 1979 inclusion map (artificial feeding)	130
8.7	Detail from 1981 inclusion map (artificial feeding)	132
8.8	Proximity map, 1979	133
8.9	Proximity map, 1981	134
8.10	Feeding and nutrition cluster, 1979	134
8.11	Feeding and nutrition cluster, 1981	135

8.12 Pathology and diseases cluster, 1979 136
8.13 Pathology and diseases cluster, 1981 137
8.14 Breeding and cultivation cluster, 1979 138
8.15 Breeding and cultivation cluster, 1981 139
9.1 Principal research themes adopted by the DGRST 143
9.2 The position of DGRST problem areas in the general industry/university network of problematisation 150
9.3 Links between problems and central poles 158
10.1 Inclusion map, 1980, central poles and passageways 165
10.2 Inclusion map, 1981, central poles and passageways 167
10.3 Zoom on 'enzyme', 1980 169
10.4 Zoom on 'enzyme', 1981 172
10.5 Proximity map, 1980, (a) first component, (b) second component, (c) third component, (d) fourth component 174-5
10.6 Proximity map, 1981, (a) first component, (b) second component, (c) third component 177
11.1 Detailed graph and skeleton graph of part of the dietary fibre graph 194
11.2 Stability of skeleton when threshold of *I* is lowered 196
11.3 How an additional link (CD) decreases the total number of links in the skeleton graph 197
11.4 Skeleton graph of the dietary fibre file, 1973–8 200
11.5 Skeleton graph of dietary fibre file, re-indexed to reduce variety of descriptors 202
11.6 Elementary patterns of key-word networks 204
11.7 Laboratories in the dietary fibre field 208

List of Tables

4.1		56
9.1	The distribution of interests identified in the proposals submitted to the DGRST	152
9.2	Dominant centres of interest in three documentary sources	154
9.3	The role played by different partners of the DGRST Concerted Action Programme in positioning research problems with respect to general socio-economic objectives	159
10.1	Patents indexed in 1981 by at least two of the three words: 'glycerol', 'triglyceride', 'enzymatic method'	178
10.2	Patents indexed in 1981 by 'submerged culture'	180-1
10.3	Patents indexed in 1980 by at least one of the four words: 'actinomycete', 'streptomyces', 'molecular structure' (*structure moléculaire*) 'anti-tumoral' (*anticancereux*)	182-3
10.4	List of key-words with frequencies in brackets	185
10.5	Key-word contents of Merck patents	186
11.1	Correspondence analysis: dimensions that account for 50% of variance	191
11.2	Number of key-words and links constituting the skeleton when threshold of *I* is lowered	195
11.3	Sorting articles by mediator words	206
11.4	Example of an article exhibiting a rare combination of mediator words	207
11.5	Activity profile of countries, measured by number of articles falling under different mediator words (in percentages of number of classified articles per country)	209

Notes on the Contributors

Françoise Bastide is engaged as researcher with the CNRS (National Scientific Research Centre) in Paris as a biologist who has recycled into semiotics. She has written several 'literary' analyses of original scientific texts in the publications of the 'Groupe de Recherches Sémio-Linguistiques', a multidisciplinary structure interested in a sign theory able to account for the production of meaning in written texts as well as in images. She has also written a study on scientific argument in *Études françaises* (with Bruno Latour) and some results on the role of illustrations in the presentation of data in *Culture Technique*.

Serge Bauin took his engineer's degree at the École des Mines, Paris and also has a Social Science degree. He has worked with the Centre of Sociology team at the École des Mines de Paris for the past five years. After field work devoted to the study of aquaculture, he is currently engaged in the mathematical development and computerisation of co-word analysis.

Michel Callon is Professor of Sociology at the École des Mines de Paris and Director of the Centre de Sociologie de l'Innovation. He is the author of numerous articles on the sociology of science and technology and science policy. He has undertaken both quantitative and qualitative research on the formation of economic markets and was directly involved in the development of policy on research in France in the early 1980s.

Jean-Pierre Courtial trained as an engineer at the École Centrale des Arts et Manufactures, has a degree in Mathematical Statistics and a doctorate in Social Psychology. He has worked both as a management adviser for news agencies as well as in the field of statistical analysis of data relating to mass media and public opinion polls. His book *La Communication Piègée* (*Communication in a Trap*) analyses the logical paradoxes inherent in communication and particularly in the mass media. His work at the Centre of Sociology at the École des Mines de Paris has led him, over the past five years, to develop a statistical and mathematical tool, corresponding to co-word analysis, and which produces a series of programmes called Leximappe.

Bruno Latour is associate professor at the Centre of Sociology of Innovation at the École des Mines de Paris and works on the anthropology of French science and technology. He has written *Laboratory Life, the Construction of Scientific Facts* (with Steve Woolgar) (Princeton University Press, new edition 1986), *The Pasteurization of French Society* (Harvard University Press, 1986), and *Science in Action: How to Follow Scientists and Engineers through Society* (Open University Press, 1986). His other works include studies in semiotics of scientific texts (with Françoise Bastide) and of scientific images (*Les Vues de l'Esprit*, special issue of *Culture Technique*, June 1985).

John Law is Senior Lecturer and Head of Department in the Department of Sociology at the University of Keele. He has written extensively on the Sociology of science, knowledge and technology. He is the joint author of *Science for Social Scientists* and the editor of *Power, Action and Belief: a New Sociology of Knowledge?* His current interests include the sociology of power and social control, in particular as these relate to machines and devices.

Arie Rip is Professor of Science Dynamics at the University of Amsterdam. He is interested in linking the philosophy and sociology of science with science policy studies and issues of science and society, and has written widely on these subjects. He is the author of *Wetenschap als Mensenwerk* (an introduction to science studies) and joint editor of *Macht over Kennis* (a volume of studies applying science studies to science policy). His current work includes studies of biotechnology and risk assessment, and involves consultancy to Dutch government departments.

William Turner is director of the Service for the Study and Realization of Advanced Information Products (SERPIA) at the Scientific and Technical Documentation Centre (CDST) of the Centre National de la Recherche Scientifique (National Scientific Research Centre). After obtaining his degree at the University of British Columbia, he went on to graduate from the Institut des Hautes Études Européennes in Strasbourg and also received a doctorate at the University of Strasbourg. He has worked as scientific adviser to European institutes in the field of the treatment and exploitation of computerised databases, and was director of CERDANE (Centre d'Études et de Recherches Documentaires sur les Attitudes Nationales vis à vis de l'Europe), at the Institut des Hautes Études Européennes. Since 1978 he has worked jointly with the Centre of Sociology at the École des Mines, Paris in the setting up and development of co-word analysis.

Acknowledgements

The work described in this volume is the product of long-term collaboration between members of the Centre de Sociologie de l'Innovation of the École Nationale Supérieure des Mines of Paris, the Service d'Études et Réalisation des Produits d'Informations Avancées of the Centre de Documentation des Sciences et Techniques (CDST) in Paris, the Department of Sociology and Social Anthropology of the University of Keele, and the Chemistry and Society Programme of the University of Leiden, and now the Department of Science Dynamics of the University of Amsterdam. As is often the case in such collaborations, individual papers written by members of one or other of these groups often owe a great deal to other members who are not co-authors. In order to avoid a plethora of mutual intra-collaboration acknowledgements we have restricted these to a few particularly important cases and tried to spell out our considerable indebtedness to those who nowhere appear in the list of authors.

Our thanks are first and foremost due to the various French Institutions which have continuously supported the lengthy research that has led to the development of co-word analysis. In particular, Jacques Sevin who was responsible at the time for the designation of Délégation Générale à la Recherche Scientifique et Technique (DGRST) research priorities, made it possible in 1980 for the Centre de Sociologie de l'Innovation to undertake an extensive programme of research on the method. This book represents one of the first fruits of this effort. The Centre National de la Recherche Scientific (CNRS) has, on several occasions, made funds available to assist the research; Edmond Lisle and Dominique Wolton have supported the work through the STS programme; Jacques Michel, former director of the CDST played a decisive role in creating SERPIA (Service d'Études et de Réalisation des Produits d'Information Avancées); G. Delacotte, Director of l'Information Scientifique et Technique of the CNRS has consistently supported our efforts. MIDIST has made funds available and encouraged us throughout. Without the resources put at our disposal by the École des Mines de Paris and its successive Directors, Pierre Laffitte and Jacques Lévy, none of this work would have been possible.

In addition we wish to thank: Jean-Pierre Courtial who, over a number of years, has brought into being and fine-tuned a methodological tool in the form of co-word analysis that is particularly well adapted to the theoretical approach of the sociology of translation; William Turner, without whose untiring activity the method would not have been developed on its present considerable scale; and Maya Sigogneau whom we have all found to be a particularly rewarding and effective collaborator.

We are grateful to the 'Chinatown' biochemists for their encouragement and help with, and to Professor Lucien Karpik for reading and commenting on, Chapters 3 and 5. We would also like to thank Dr Rob Williams of the Department of Clinical Health Studies of the South Glamorgan Institute of Higher Education for permission to reproduce material in Chapter 5.

We are particularly grateful to Serge Bauin and Jean-Pierre Courtial, both of the Centre de Sociologie de l'Innovation of the École Nationale Supérieure des Mines de Paris for allowing us to use material gathered from their work on aquaculture and dietary fibre, assisting us in the interpretation of Leximappes, and commenting on early drafts of Chapter 7.

The research reported in Chapter 9 was jointly financed by the Délégation Générale à la Recherche Scientifique et Technique and by the Centre National de la Recherche Scientifique. We would especially like to thank Mr Guy Meyer, researcher at the Strasbourg School of High Polymers, and several members of the Scientific and Technical Documentation Centre of the CNRS responsible for macromolecular chemistry, G. Zeltzer, J. Rauss and M. Rosenburg, for their very helpful comments on an initial version of Chapter 9.

We would like to thank M. Minoux, Chef du Département Informatique et Automatique de l'École Nationale Supérieure des Techniques Avancées, Conseiller Scientifique du Centre National d'Études des Télécommunications for his contribution to the development of the methods described in Chapter 11. We are also grateful to P. O. Flavigny and José Juan, methodologists and computer scientists at, respectively, the Departement de Mathématiques Appliquées du Centre d'Études Sociologiques, and the Centre de Documentation Scientifique et Technique of the CNRS for writing the computer programmes described in this chapter, and to Dr John Whittaker of the Department of Sociology and Social Anthropology for reading and commenting on an early draft.

We wish to thank the University of Keele, the École Nationale Supérieure des Mines, la Fondation Fyssen, the Leverhulme Foundation, the Economic and Social Research Council and the CNRS which made possible the collaboration between members of the Department of Sociology and Social Anthropology at the University of Keele and members of the Centre de Sociologie de l'Innovation of the École Nationale Supérieure des Mines de Paris.

Glossary

Actor-network
The structure and operation of an actor-world (q.v.): an interrelated set of entities that have been successfully translated (see translation) or enrolled (see enrolment) by an actor that is thereby able to borrow their force and speak or act on their behalf or with their support. The entities may be seen as forming a network of simplified points (see also black box) whose simplicity is maintained by virtue of the fact that they are juxtaposed with others. The actor who speaks or acts with the support of these others also forms a part of the network (see translation centre). Hence the term actor-network, for the actor is both the network and a point therein. It should also be noted that each point entity that is enrolled in an actor-network depends on its capacity to translate other actor-network (q.v.). A simplified entity that is nevertheless also a network in its own right.

Actor-world
The world of entities generated by an actor-network (q.v.). Emphasises that, for any given actor, there is nothing beyond the network which it has created, which constitutes it, and of which it forms a part.

Black box
A way of talking of the simplified points that are linked together in an actor-network (q.v.). A simplified entity that is nevertheless also a network in its own right.

Enrolment
The definition and distribution of roles by an actor-world (q.v.). It should be noted that roles are not fixed and pre-established, and neither are they necessarily successfully imposed upon others.

Funnel of interests
A series of problematisations (q.v.) that starts with a problem (or

interest) that is generally held to be important and then translates the reader/hearer to a specific problem-solution. Typically found in the introduction to scientific papers.

Interessement
The action of interesting, enrolling or translating which involves one entity attracting a second by coming between that entity and a third. Interessement is thus a transaction between three entities. It may be seen as the elementary form of translation which, however, should not be confused with the operation of social interests.

Problematisation
Most generally, a form of translation (q.v.) that posits an equivalence between two problems that requires those who wish to solve one to accept a proposed solution for the other. Accordingly, a form of control that involves imposing an itinerary upon others by making oneself indispensable to them. In science this itinerary usually starts with a general problem which can (if the problematisation is successfully imposed) only be resolved by the solution of a much more specific problem proposed by a particular laboratory (see funnel of interests). In other words, scientific problematisations usually impose a detour through a specific research site.

Translation
The methods by which an actor enrols others. These methods involve: (a) the definition of roles, their distribution, and the delineation of a scenario; (b) the strategies in which an actor-world (q.v.) renders itself indispensable to others by creating a geography of obligatory passage points; and (c) the displacement imposed upon others as they are forced to follow the itinerary that has been imposed. The elementary form of translation is that of interessement (q.v.). A common form of translation in science is that of problematisation (q.v.).

Translation centres
Locations at which translation strategies are evolved, attempts are made to control the diverse elements that make up the actor-network (q.v.), and calculations about the return from different strategies are made. Locations which are able to turn themselves into spokesmen for other entities and, in some sense, profit from this asymmetry.

Part I
Introduction

1 How to Study the Force of Science

MICHEL CALLON, JOHN LAW and ARIE RIP

SCIENCE IN SOCIETY

Let us start with a statement of the obvious: science and technology are powerful forces in modern industrialised society and are accordingly of vital direct and indirect importance to many. Of course they are of central interest to scientists and science policy-makers, but they also concern parties as diverse as big industry, the military, governments, lobbies, groups of concerned citizens and the general public which may be both excited by and feel powerless in the face of scientific advance. At the same time, however, their inner workings remain opaque. The way in which their force is created and deployed is obscured by the wide range of ways in which this can occur, and by the myth-making that surrounds the whole process.

In this book we present an approach which is designed to clear such opacity. This approach permits a description of the way in which the force of science is put together (and taken apart) in the everyday activities of scientists. It also makes it possible to depict the development of domains of science by means of a method called co-word analysis. The book is thus a contribution to the study of the dynamics of science, and will be of interest to scholars in that field, to science policy-makers and, it is hoped, to scientists themselves. However, there is also a wider claim: that an understanding of the dynamics of science is only possible when the force of science in present-day societies is taken into account. Our way of starting on this task is not to present a full-blown analysis of society and social change. Rather it is to adopt a method that does not distinguish on *a priori* grounds between 'science' (which is purportedly about the 'truth') and 'politics' (which supposedly concerns 'power'). It is our argument that a proper understanding of social and scientific change requires the abandonment of this dichotomoy.

Touraine, in one of his historical panoramas (Touraine, 1973), has attempted to identify the strategic loci in society, the hubs of change where developments are shaped and society is transformed. After the medieval monastery and cathedral, the palace became the strategic locus during the Renaissance. Only after the first phase of the Industrial Revolution was this replaced by the industrial firm, while now, according to Touraine, it is the laboratory that has, in turn, displaced the industrial firm. The argument, then, is that as a class, laboratories currently exercise crucial social power.[1] At present it is difficult to enter into political or social struggle, or issue a challenge that might transform society, without seeking at least some support from science, its achievements and its social status. In their attempts to exert some force on society, actors typically pass through the laboratory: they wield the scientific or the technological in their attempts to transform society. Our argument, documented in the chapters that follow, is thus, that to understand our societies and their transformations we have to follow the actors and uncover the fact that not only palaces but also laboratories lie along their paths. Our study of social structure and social change thus follows Touraine's injunction to study the strategic locus of the laboratory.

Our central methodological prescription has already been stated: it is to 'follow the actors'[2] both as they attempt to transform society and as they seek to build scientific knowledge or technological systems. But to put it this way is already misleading. In particular, it is vital not to be diverted by the myth that says that there is a gap between science and politics and that the two are, or should be, separate. Our argument which follows that of Latour (1983, p. 168) is that science is politics by other means and, accordingly, that the study of science takes us straight into politics. For outsiders science and technology may take the role either of a demon or a god (Weber, 1948), while insiders mobilise both the 'scientific' and the 'political' in their endless struggles. The idea that there is a special scientific method, a realm where truth prospers in the absence of power, is a myth. Indeed, it is particularly important to follow actors closely when they enter strategic loci, for it is often in the interests of the forces at work to conceal the way in which they act.

THE PRINCE AND THE SCHOLAR

The thoroughgoing determination to follow the actors wherever they go, to uncover what they might prefer to keep concealed, and to avoid being misled by myths is nowhere better exemplified than in the writings of Machiavelli. In his famous study, *The Prince* (Machiavelli, 1961), he reveals the way in which power was exercised in the strategic locus of the 'palace' in Renaissance society. His approach, and in particular the calm and even-handed audacity with which he entered the palace and observed what went on, serves as an admirable model for our own study of the laboratory. Thus, he describes in a matter-of-fact way the actions of Cesare Borgia, his chosen 'prince', in a way that underlines the cynical but effective duplicity of the latter. For instance, Borgia engages the services of a mercenary without scruple in order to suppress a popular revolt, but later offers the same mercenary as the perfect scapegoat to those he has suppressed. The lesson to be drawn from this and other examples has become proverbial: political efficacy cannot concern itself overmuch with the principled precepts of moral law.

In order to show the duplicity of princes, audacity is needed. It is no good drawing idealised portraits. Machiavelli pays for his audacity with his liberty, and only narrowly escapes with his life. In this sense there is no contrast between the republican enthusiasm of the *Discourses* (Machiavelli, 1950) and the lessons in amorality offered in *The Prince*. Political theorists have long struggled with the problem of how machiavellian Machiavelli intended to be. There is no simple answer to this question, and it is certainly not one we wish to become involved with here. What is important for us is rather Machiavelli's method. In a detailed and systematic description of the machinations of princes which flushes out their hidden designs, he devotes himself, as a participant observer, to what would nowadays be called an ethnography of political action. Establishing his quarters in the palaces of Florence, and following the prince, Machiavelli succeeds in bringing Italian society to life and understanding its history and its conflicts. Paradoxically, in attaching himself to one person, the prince, he reconstitutes the cruel reality of a whole society. To be rigorous in such a project it is necessary to be audacious. In particular, it is vital not to let morality blind oneself in the study of how a society takes shape and is transformed from its strategic loci.

The location has changed but the method remains the same: describe, with neither fear nor favour, what it is the actors do. For the idealised prince Machiavelli substitutes the calculating prince, and by this simple displacement he makes the efficacy of the prince comprehensible. In the same way, the study of the daily reality of scientific work has to replace second- or third-hand stories and idealisations. Ethnographers of the scientific laboratory have been producing such accounts and it is time for us to use them to understand the force of science.

As we follow the actors of science there is no taking sides, no question of charging science with all the sins of the world or pleading its complete innocence. This is not our intention, and neither should it be. Again we cite Machiavelli as our example. He does not intend to stigmatise the behaviour of the prince by revealing his villainy to a horrified world. As Aron points out in his Preface to the French edition of *The Prince* (Aron, 1972), Machiavelli's sympathetic analysis of despotism is linked with his attachment to liberty. Such a realism even if it entails violation and usurpation, appears inevitable to the careful observer as soon as political action establishes peace and the continued reign of a civilising state. In a famous passage from *The Prince*, Machiavelli notes, '. . . I believe that it is probably true that fortune is the arbiter of half the things we do, leaving the other half or so to be controlled by ourselves' (Machiavelli, 1961, p. 130). Without the art of the Prince, he continues, Fortune, like swollen rivers, sweeps everything before it. 'Yet although such is their nature, it does not follow that when they are flowing quietly one cannot take precautions, constructing dykes and embankments . . .' (Machiavelli, 1961, p. 130).

Is it not the same with the science and technology of our own time? Without them human action is dominated by immediate necessity and has little power. The effective exercise of choice requires the support of science and technology, even if they simultaneously contribute to the creation of constraints on action. Paradoxically, they have become necessary even to those societies and groups that would like to escape from their increasing dominance. Science is inevitable and indispensable as is usually recognised by those who are dominated by it. Thus there is no question of rejecting science, or of joining a vaguely Rousseauist or openly irrationalist movement. Rather, the problem is to understand, through careful study, the roots of scientific power, and where this is taking us.

There is, therefore, no need to take sides. Machiavelli's analysis serves the prince just as much as it serves his subjects. There is no need to follow Rousseau who takes *The Prince* to be an object-lesson for republicans (Rousseau, 1968). Why should we think of Machiavelli as a cunning impostor who poses as an adviser to the prince in order to expose him? A non-partisan analysis may be corrosive and have the effect of destroying the status of its object, but its primary aim remains that of making political action less opaque and possibly more efficacious. The same argument holds for the study of science and technology.

Now consider another lesson offered to us by Machiavelli – his anti-reductionism. Is it not tempting to reduce the evolution of societies to the actions of a few princes who exert their power? Or, alternatively, to consider princes as puppets without substance, whose strings may be pulled by such powerful social actors as the people or the nobles? Machiavelli avoids both temptations. He shows us a prince who is skilful in the art of managing variable and unexpected social forces. The prince is not a lonely man, drawing force from his tactical genius alone, yet neither is he the simple plaything of forces that transcend and manipulate him.

Students of science and technology have not always been so wise. They have often been tempted either by sociological reductionism or by its converse, an idealisation of scientific genius or method. Thus much (though by no means all) of the philosophy of science celebrates the view that science is an activity apart, pursued by disinterested intellects who have special access to methods for determining the nature of the world. The same image is to be found in certain sociological writings on the institution of science. Thus Merton, who may be seen as the founder of the sociology of science in the Anglo-Saxon world, pictured the activity of science as one protected from the normal, often self-serving, interests of other sectors of society by a barrier of insulating norms (Merton, 1973). And these views are not limited to specialists in the study of science. The notion that science is in some way pure, set aside from daily and profane activities, is deeply entrenched in Western culture.

At the same time, however, there has been a counter-argument put forward by students of science who take the links between scientific developments and other social institutions as their starting point. Following Marx, the links have sometimes been posed in simple terms: where science is pursued in a class society the nature of scientific innovation will inevitably reflect the interests of the dominant class (see, for example, the classic Bukharin *et al.*, 1971). More sophisticated writers, inspired by Marx or Weber, hold that the very nature of science and technology is such that they are constituted as a form of domination and are thus incompatible, as they at present stand, with humanistic and egalitarian social relations (see, for instance, Habermas, 1971). Yet others argue that scientific work should be studied as similar to any other kind of work in a class-based, exploitative society (Levidov and Young, 1981).

Even if Marxist views of science have not become dominant, the general counter-argument that they represent has gained ground over the course of the last twenty years. The majority of students of science today accept that it is not possible to understand the nature and direction of scientific innovation, even in pure science, if it is treated as something that takes place in an ivory tower. A sociology of scientific knowledge is attempted (see Barnes, 1977; Bloor, 1976). Indeed, a number of outstanding historical and sociological studies have sought to tease out the relations that exist between the content of scientific knowledge and the interests of such social collectivities as professional groups and social class (see, for instance, MacKenzie, 1978).

Stimulating though such studies are, they run the risk of sociological reductionism. The content of science becomes transmuted into the simple reflection of social interests. While the 'ivory tower' view of science cannot explain why the laboratory is the locus of such powerful transforming social forces, this sociological reductionism leaves us with the alternative question: why is it that it is the *laboratory* that is the strategic locus in modern society, rather than (say) the boardroom or the trade union head office? Both views err by isolating, and in a sense black-boxing, one side of the equation. The notion of a society that can shape science without itself being influenced is as false as the converse image of a science and technology that find themselves able from their own resources to impose a structure unilaterally on their social environment. In place of these reductionisms it is best to follow another route, that of non-reductionism preferred by Machiavelli. Science and technology have to be seen as loci of strategic action in which existing scientific cum social relations are worked upon in order to produce, in one and the same movement, both new knowledge and novel social relations.

Indeed, as we have already indicated, the separation of knowledge from the social, though avoided only with difficulty for linguistic reasons, is already misleading. It is vital to avoid a dual repertoire, one to describe the social and the other to describe the content of science and technology. It is not that the scientist, a skilful strategist outside the laboratory, becomes rational and disinterested once he is within its walls. Rather, a single, undifferentiated, vocabulary is necessary if we are to follow the daily actions of scientists as they act strategically to redefine both science and society while routinely ignoring *a priori* distinctions. To do so, we should not hesitate to enter the laboratory, just as Machiavelli before us explored the recesses of the palace.

THE PRACTICE OF SCIENTISTS AND THE ROLE OF INSCRIPTIONS

A series of recent ethnographic studies of laboratory science has documented the range and diversity of the daily work of research scientists (Knorr-Cetina, 1981; Latour and Woolgar, 1979; Law and Williams, 1982; Lynch, 1985; Williams, 1984; Williams and Law, 1980). A striking feature of their work is the way in which scientists are constantly engaged in the juxtaposition of heterogeneous elements: delicate physical manipulations with rats, test tubes and radiation counters; interaction with colleagues, technicians, laboratory directors and suppliers; reading letters, trips to the library, drafting papers, revising manuscripts, going to conferences, writing grant applications – all these and other elements too numerous to mention are fitted together during the course of the typical working day. Bits and pieces are assembled in an attempt to impose coherence and ensure that the research in question does not go off course, that publishable results are generated and that the reputation of the laboratory and/or the individual is maintained.

The behaviour of the scientists studied conforms in every way to that of the classical picture of the entrepreneur. They attempt to obtain the use of various kinds of resources, to fit them together and to profit from the results. In so doing they display a general lack of concern with the distinction between what is 'internal' to science and technology, and what is 'external'. Although they may be able, on occasion, to articulate such a distinction, in their daily activities the 'scientific' and the 'non-scientific' are indiscriminately jumbled together. This does not, however, mean that their work is incoherent. Quite the opposite is the case. Scientists are building a structured world and the attempt by analysts to force it to conform with criteria of demarcation tends to distort, if not to destroy, the coherence of that world. Building a world in which experimental rats in one laboratory have a story to tell to readers of scientific journals all over the world requires the management of heterogeneous resources. The rats have to be bought and paid for, kept free from infection, enclosed in cages, prepared for drug injection and subjected to measurement. But the work of technicians must also be organised, laboratory resources must be secured in the face of competition from colleagues, research applications must be filed and papers must be written and presented.

Controlling resources, controlling the environment and controlling the world that is being built, all of these are aspects of the entrepreneurial activity of scientists. In a sense then, they are not only practising science – they are also practising politics, economics and sociology. Thus the analyst who follows scientists into their laboratories has no need to create his or her own categories and linkages. Since scientists are also practical politicians and sociologists, they are able to supply them in profusion themselves. The job of the analyst is rather to study the creation of such categories and linkages, and examine the way in which some are successfully imposed while others are not.

How, as a matter of fact, do scientists go about creating or controlling entities in their environments, and enlisting them as they build their versions of the world? In the immediate environment they can exert manual force, use machines, make use of money, talk, create or employ texts and other inscriptions and harness physical or chemical forces (for instance, electric power or chemical reagents). However, if they wish to extend their worlds beyond the physically immediate, then the means have to be found to act at a distance. The types of agents that are so mobile are relatively limited in number: bodies, machines, money and reagents are transportable, as is the human voice via the telephone. Perhaps most crucial, however, are inscriptions.

First, observe that in our society inscriptions are of general importance. From notes, memos, letters and *curriculum vitae*, to account books, cheques, newspapers, magazines, books, certificates and official documents – most of us make use of different kinds of inscriptions dozens of times each day.[3] Scientists attach especial importance to inscriptions: output from recorders, laboratory notebooks, graphs, computer printouts, preprints and published scientific articles; all of these appear routinely in the daily round of work. As will be argued in more detail in later chapters, the particular importance of texts in science lies in the fact that they constitute a central political tool for the scientist–entrepreneur: in the text a structured world is built up that encapsulates the world-building activity of the scientist and allows him/her to act on others at a distance. The handling of the experimental rats is described in a methods section, the recorded reactions to drug injection are transformed into a table and a graph in the results section, and the claims of the scientist about how these 'data' should redefine the field form a part of the discussion and conclusion. The force of such texts also resides in the fact that they contain links with other texts (references to the literature), work and institutions. These are all built into the world presented in the text and may thus be transmitted between contexts. In sum, texts make possible the construction of linkages between existing entities and the formation of novel entities and, if persuasive, thereby constitute an important method for attempting to control the environment.

Although we do not want to reduce scientific and technological activity to the transmission of literary inscription, we none the less consider inscriptions to be a central feature of science and technology. Scientists themselves attach particular importance to texts, treating them both as a goal (the publication of a scientific article) and as a means (a way to build a world and persuade others). Since they are also relatively explicit (for they have to stand on their own beyond the laboratory walls) our argument is therefore that though science cannot be reduced to texts, these nevertheless constitute a prime source of data for any study of the way in which worlds are created and transformed in the laboratory. Instead of following the actors we may therefore follow the texts. We may not, in this way, find out everything there is to know about scientists in laboratories. However, the study of texts is well suited to our particular and limited task: that of studying scientific change in order to understand the power of the laboratory.

WORLDS OF SCIENCE AND HOW TO MAP THEM

Some texts manage to convince those to whom they are sent, while others are returned to be rewritten or abandoned as failures. It follows that we have to leave the laboratory if we want to discover their ultimate fate. Within the laboratory, the talk of scientists centres on ways of ensuring the acceptance of texts: on the features of scientific journals, institutions and colleagues that might allow these to be enrolled in the world being built in the text.[4] There are no guarantees, but there is an accepted strategy. This is to construct the scientific paper by forcefully linking elements derived both from the laboratory, and the institutions and scientific research area at which the paper is aimed. The force of this political ploy by the scientist–entrepreneur resides in the way a world is built within the text that presents a story of, say, experimental rats and their reactions to the injection of drugs. The success of a given text lies in the extent to which its authors manage to enrol others such that the latter accept the picture proposed to them, offer the authors recognition and credit, cite the paper in their own publications, and make grants available to them in the future. It resides, in other words, in the extent to which the authors are granted resources that allow them to continue their work. Such success has been measured qualitatively and in detail by the laboratory ethnographers and more quantitatively (and often in an aggregated way) by those concerned with providing data as a basis for policy decisions (see Elkana *et al.*, 1978). There are many problems, both practical and theoretical, in the measurement of achievement in science. However, our concern is not to offer a method for scoring relative success. Rather it is to present a method for surveying the dynamics of science as these are constituted by contributions from laboratories and their mode of interaction.

It would, perhaps, be possible to stretch the ethnographic method to study the structure of research areas (similar broad ethnographies are not uncommon in anthropology). However, the difficulty of such an approach is obvious because domains of science are non-localised in character. It would, in addition, be very time-consuming. But there is no need to undertake ethnography if it is possible to study the fate of scientific actors and their worlds by studying texts. The scientific paper achieves its success through the (partial or full) acceptance by others of the structure of entities ('scientific' or otherwise) proposed in the paper. If a paper linking cancer chemotherapy with a polymer, DIVEMA, (see Chapter 5) is accepted for publication, and is cited after publication, then this counts as a success for its authors. The success of their world-building is, however (and this is the crucial point), visible in other texts that make similar links. Cancer chemotherapy and DIVEMA now go together. Other scientists (or such third parties as policy-makers) who wish to build a world of their own and thus enrol and control their environments, have no alternative but to take this into account. In other words, this particular network of linkages becomes available as a resource. In this way a world of science is built up, not through the juxtaposition of elements by individual actors, but rather as an emergent side-effect of interaction between the laboratory strategies of a wide range of actors. Or, to put it in terminology derived from a different sociological perspective, we see the unintended consequences of action at the collective level.

A scientific text not only reveals the world-building strategy of its authors, but also the nature and force of the building blocks derived from the domain of science from which it draws and to which it contributes. The text thus provides access to the dynamics of science, to the shared worlds that constitute a means of mutual (and evolving) control. But how can such worlds of science be mapped? In principle, the answer is simple. The text may be treated as a structure linking important words. Thus, in the paper discussed in detail in Chapter 5, 'therapy' is linked to 'DIVEMA' with the help of the physiological phenomenon 'pinocytosis'. The whole cluster is linked to 'tumour', a word important to the readership of the journal to which the paper has been submitted. If the linkages are accepted and appear elsewhere, this cluster of terms becomes a durable feature of research on cancer chemotherapy.

By following the texts the analyst may trace the appearance and disappearance of forceful words and durable linkages and so build up a picture of the area of science in question. Of course, this has always been part of the historian's craft.[5] If it is more specified in the present context, this is only because the analytical perspective has been made more explicit. Our exclusive focus on important words certainly loses some of the fine-grained structure of scientific papers that concerns historians and scientists. But there is a corresponding gain: converting texts into skeletons of words makes it possible both to handle larger databases, and to use statistical and graphical methods for the display of features of the worlds of science. In short, such methods solve the problem of aggregation, of identifying the intended and unintended consequences of scientific action at the collective level without having to assume the perspective of one of the participants.

The third part of this book sets out the method of mapping the worlds of science in detail and presents a number of examples. Since the conversion of many texts into a short series of important words is very laborious, the indexes of articles, reports and patents prepared by scientific and technical documentation services are used. The latter have their own objectives (for instance, the facilitation of literature searches) and their simplifications introduce error: specifically, there is a loss in immediacy and accuracy which leads to difficulty in interpreting the maps. On the other hand, the use of such indexes makes it possible to produce maps on a routine basis from large, computerised databases which cover wide areas of science.

The advantages of using computerised scientific databases to construct maps and analyse science and the activities of scientists have been recognised and exploited during the last ten to fifteen years. Indeed, a new speciality, that of scientometrics, has grown up within the domain of the social studies of science and technology.[6] Though our method falls into the scientometric tradition, it differs from such other approaches as co-citation analysis because it is built around important words within scientific articles. This means that it is possible to make a direct link between the strategic world-building of individual scientists and the collective outcome of such activity. We have, in other words, sought to pursue the qualitative by other means.

CONCLUSION

In the preceding sections we have shown in outline how the force of science may be analysed. Our argument is that a description of the deployment of force by assembling and linking entities, be they rats and graphs, scientific institutions, or deadly diseases, must be central to any study of the dynamics of science. In the third section we briefly discussed the way in which actors juxtapose entities and considered the importance of texts in such a process. In the last section we followed the texts on their journey out of the laboratory, discussed their fate, and considered the way in which they interact to produce emergent effects at a collective level. The worlds built by scientific entrepreneurs may enrol others and endure, or they may fail to do so and leave no trace. The production and distribution of texts that encapsulate, as it were in a microcosm, the world in which the reader is being asked to enrol, constitutes a vital method for building worlds in science. Such texts may be analysed in order to uncover the traces of enduring and successful worlds. In fact, the structure of a scientific domain now appears as the network of traces to be taken into account by any new text that aims to contribute to that domain.

To follow actors and texts wherever they go with neither passion nor prejudice, this is a powerful method for describing the dynamics of science without reducing these to the extremes of either internalism or externalism. However it does not and cannot *explain* the success or failure of strategic world-building in the strong sense of prediction. This is because there is no guarantee of success, even to those who have succeeded before. The labours by which societies give birth to their history are unpredictable.

It is, however, possible to do away with the opacity that surrounds science and technology. There are long chains that link such conventionally political entities as government bodies, industrial enterprises, hospitals and voluntary associations with such covertly political entities as polymers, experimental rats, electrodes or fuel cells (see Chapter 2). *All* of these are political because they are potentially capable of exercising force, but the political character of some is concealed by the myths that disconnect science from what normally counts as politics. By entering laboratories and following actors we are able to make the connection and so render transparent a political debate on science and technology that is currently lamentably uninformed. Scientists and technologists conduct politics by other means. And in contemporary societies the laboratory is a strategic locus: science and technology are hubs of social transformation and thus inescapably political in nature – a fact that becomes very clear once the co-word findings are considered.

The act of entering the laboratory and coolly describing what one sees does not imply that sides have to be taken for or against science and technology, even if such descriptions tend to be corrosive of established myths. As Machiavelli's work shows, description serves to enhance the efficacy of *all* parties: scientists, technologists, policy-makers, pressure groups; all may profit from it. To repeat, a 'political' analysis of science and technology should not be seen as a covert way of accusing the practitioners in these disciplines of duplicity. We are not in the business of exposing a set of machinations that are known to and well recognised by insiders. Scientists and technologists themselves also suffer from ignorance in the face of the politics of science for they only know their little corner of the game. If our approach counts as an attack on anything at all, then it is perhaps at the reductionist traditions of the study of science and technology that it is aimed. For too long scholars in this field have been content to simplify the heterogeneous links that tie the 'scientific' and the 'technological' to the rest of the social world. Sometimes, as we noted earlier, these institutions are seen as a direct reflection of social forces. On other occasions they are held to have no relationship with the latter at all. If this volume achieves nothing else, we hope that the reader will be persuaded of the essential continuity between science and society, and of the fact that similar methods may be used in their analysis.

NOTES

1. Here we use the 'laboratory' as a generic term to cover all research sites.
2. Our conception of 'actor' is broader than that which is normal in Anglo-Saxon sociology. See later and particularly Chapter 2.
3. For complementary anthropological and historical perspectives on the importance of literacy and the printing press see Goody (1977), Eisenstein (1979) and Latour (1985).
4. For a study that concentrates on such strategic decision-making see Law and Williams (1982).
5. See Mackenzie (1978) and, for a more 'internal' example, Mullins (1972).
6. Work in this speciality is reported in *Scientometrics*. See also Garfield, Malin and Small (1978); Griffith, Small, Stonehill and Dey (1974); Small (1977); Small and Greenlee (1980a, 1980b); Small and Griffith (1974).

Part II
The Power of Texts in Science and Technology

2 The Sociology of an Actor-Network: The Case of the Electric Vehicle

MICHEL CALLON

INTRODUCTION

If laboratories and research sites are to the twentieth century what monasteries were to the twelfth, then the sources of their power and efficacy remain a mystery. How is it that the ideas and writings that issue from these institutions are able to revolutionise, if only gradually, conditions of work in industry, the universe of consumer goods and lifestyles? How are the discoveries made in Stanford, Gif-sur-Yvette, and Cambridge diffused such that they become universally known and recognised? How are certain technical devices, shaped in research departments of French or English companies, able to conquer markets throughout the world? Anthropological studies of the laboratory have shown that nothing exceptional occurs within the walls of research centres themselves which could account for their influence. These studies have also shown that the force and generality of results obtained cannot be attributed to the existence of a specific scientific method (Latour and Woolgar, 1979; Knorr-Cetina, 1981; Lynch, 1985). Though scientists give certain activities a higher priority than others (see Chapter 3), the former do not possess greater rigour or a logic which enable an observer to distinguish them from the latter. In addition, the study of controversies has discredited the view that science and technology are free from any influence other than the intellectual (Collins and Pinch, 1982; Shapin, 1979; MacKenzie, 1978). Rather, they are impure and heterogeneous. If the hesitations, changes and evolutions that mark their development are to be understood, then interests, strategies, and power relationships which do not stop at the laboratory door must also be brought within the scope of analysis. In sum, though science and technology develop in some measure apart from the rest of the world, they are neither detached nor fundamentally different in nature from other activities.

The power of science must thus be explained in another manner. Yet though sociology and anthropology have played a decisive role by describing the detailed content of scientific practices and have undermined a range of classical assumptions about science, they have unfortunately failed to account satisfactorily for its undeniable influence. This is because they have sought to explain its origins and success in terms of supporting political interests (MacKenzie, 1978), resources gathered by researchers (Latour and Woolgar, 1979; Knorr-Cetina, 1981), or pressures of economic demand (Hessen, 1971, Yoxen, 1981). In short, they have searched for the causes of scientific power not within science but rather within the surrounding society. Unfortunately, these explanations are inadequate for, as is shown by an increasing number of studies, the development of scientific knowledge and technical systems cannot be understood unless the simultaneous reconstruction of the social contexts of which they form a part is also studied (Latour, 1984; Callon, 1980b, 1981a; Law, 1985b; Pinch and Bijker, 1984).

In order to uncover the mechanisms of power of science and technology, it is therefore important to reveal the ways in which laboratories simultaneously rebuild and link the social and natural contexts upon which they act. In this chapter we present three concepts which make it possible to analyse this 'co-evolution' of 'society', technological artefacts and knowledge of nature. The three concepts are those of the actor-world, translation and the actor-network.

ACTOR-WORLDS

Let us start, then, with a short detour, by posing a question that is of perennial interest to sociologists: of what and of whom is society made up? Sociologists often believe that, as a profession, they are better able to answer this question than others. Are they not social specialists who know how to juggle groups, classes, orientations, habitus and power in order to explain and predict behaviour?

However, scientists and technologists constantly undermine the knowledge and competence of the sociologists. This is because they are continuously rebuilding society by introducing unpredictable variations and new associations. Furthermore, it is from these variations and associations that they draw their power. More than any other kind of actor, technologists may be sometimes endowed with the capacity to construct a world, their world, to define its constituent elements, and to provide for it a time, a space, and a history. (Latour, 1984; Hughes, 1983). To illustrate this ability we will follow the development of a technological innovation that was thought to be of major importance in France during the early 1970s: that of the electric vehicle, the VEL (véhicule électrique).

In 1973, the EDF (Electricité de France) presented a plan for the VEL that not only determined the precise characteristics of the vehicle it wished to promote, but also the social universe in which the vehicle would function.

First, the EDF defines a certain history by depicting a society of urban, post-industrial consumers grappling with new social movements. The motor car occupies a highly exposed position, for it forms part of the world that is under attack. Thus it serves as a point of departure for the construction of far-reaching and radical demands which will lead to a future that may be discerned only with difficulty. The internal combustion engine is the offspring of an industrial civilisation that is behind us. The Carnot cycle and its deplorable by-products are stigmatised in order to demonstrate the necessity for other forms of energy conversion. On the one hand, the motor vehicle is considered responsible for the air pollution and noise that plague our cities. On the other hand, it is irretrievably linked to a consumer society in which the private car constitutes a primordial element of status. However, electric propulsion will render the car commonplace by decreasing its performance and reducing it to a simple, useful object. The electric car could lead to a new era in public transport in the hands of new social groups which are struggling to improve conditions in the city by means of science and technology. The goal would be to put science and technology at the service of the user, and to do away with social categories that attempt to distinguish themselves by their styles of consumption. The EDF has based this vision on an evaluation of the trajectories of development open to different types of electrochemical batteries.[1] Firstly, public transport could be equipped with improved lead accumulators. Then, accumulators and fuel cells could open up the larger market of private transport by enabling the VEL to reach speeds of up to 90 km/h.

By predicting the disappearance of the internal combustion engine due to the rise of electrochemical generators and by ignoring traditional consumers so as to better satisfy users who have new demands, the EDF has not only defined a social and technological history but also identified the manufacturers that would be responsible for the construction of the new VEL. The CGE (Compagnie Générale d'Electricité) would be asked to develop the electric motor and the second generation of batteries, and to perfect the lead accumulators to be used in the first generation of the VEL. Renault would mobilise its expertise in the production of traditional automobiles in order to assemble the chassis and make the car bodies. The government would also be enlisted: this ministry would formulate regulations favourable to the VEL; that ministry would subsidise those municipalities interested in electric traction. The list goes on: companies that run urban transport systems would cooperate with research centres, scientists, etc. The EDF has defined the roles, and then attempts to enrol other entities into them.[2] It binds the functions of these roles together by building a world where everyone has his own place.

Up to this point, the entities are ones familiar to the sociologist. There are consumers, social movements and ministries. But it would be wrong to limit the inventory. There are also accumulators, fuel cells, electrodes, electrons, catalysts and electrolytes. For, if the electrons do not play their part or the catalysts become contaminated, the result would be no less disastrous than if the users rejected the new vehicle, the new regulations were not enforced, or Renault stubbornly decided to develop the R5. In the world defined and built by the EDF, at least three new and essential entities must be added: zinc/air accumulators, lead accumulators, and fuel cells with their cohort of associated elements (catalysts, electrons, etc.).

But just a moment. EDF is not an actor that is faced with new and unfamiliar technologies or knowledge located throughout society. Neither is it an imaginary construction that might be considered as unrealistic by an experienced sociologist. Nor is it a simple world. There is what we propose to call an actor-world, a world of which EDF, its prime mover, forms a part. EDF puts forward a list of entities and a list of what they do, think, want and experience. These entities are not human alone since electrons, catalysts, electrolytes and lead accumulators are included. These entities act, react and cancel each other out, in just the same way as any others. They may be either individual or collective. Actor-worlds have no favourites and are racialist only by omission.

The actor-world not only determines the repertoire of entities that it enlists and the histories in which they take part. It also determines their relative size. For the EDF actor-world, Renault is no longer a powerful company that seeks to be the largest European car manufacturer.

Indeed, it will never regain that status. Rather, it is reduced to the level of a modest entity that intervenes in the assembly of the VEL. The same is true of the status groups that give way to social movements and their new demands.

The notion of the actor-world makes it possible to describe the contents of technical objects and theoretical knowledge. The VEL is one of the results of the identification and interpretation of entities by the EDF actor-world. The ingredients of the VEL are the electrons that jump effortlessly between electrodes; the consumers who reject the symbol of the motor car and who are ready to invest in public transport; the Ministry of the Quality of Life which imposes regulations about the level of acceptable noise pollution; Renault which accepts the fact that it will be turned into a manufacturer of car bodies; lead accumulators whose performance has been improved; and post-industrial society which is on its way. None of these ingredients can be placed in a hierarchy, or be distinguished according to its nature. The activist in favour of public transport is just as important as lead accumulators which may be recharged several hundred times.

In the absence of one ingredient the whole would break down. If the user were removed this would be just as serious as a failure to ensure the durability of the fuel cells. In both cases the vehicle would stop dead. In other words, the existence of the VEL is bound up with the construction of the actor-world. This particular technical object cannot be understood without considering the fact of its concomitant actor-world. It does not matter whether it is the catalyst that gets contaminated, the consumer who becomes dissatisfied with poor performance, or Renault which refuses to fall into line; if any of these happen the VEL will cease to work and exist. Accordingly, technical objects must be seen as a result of the shaping of many associated and heterogeneous elements. They will be as durable as these associations, neither more nor less. Therefore, we cannot describe technical objects without describing the actor-worlds that shape them in all their diversity and scope. The VEL is just like the actor-world that both supports and is being supported by it. It is scientific, political, and economic throughout. This is because it is a combination of elements borrowed from these different registers which the sociologist claims to be able to distinguish from one another.

Once we accept that the VEL is constructed from this range of heterogeneous elements, the next question to ask is how we can describe its construction. As we have just seen, this is the same as asking how our actor-world is built, for without the actor-world, the technical object would not exist.

TRANSLATION

An actor-world associates heterogeneous entities. It defines their identity, the roles they should play, the nature of the bonds that unite them, their respective sizes and the history in which they participate. But actor worlds must not be represented as shoppers in a well-stocked supermarket choosing what they wish to buy from a pre-established list. Once an actor-world comes into being, it does not draw its entities from previously established stock. It is not constituted in the way a shopping cart is filled. In short, there is no world, or worlds, from which pre-existing elements can be extracted. Nor is there a world which guarantees that the combinations created by the actor-world are realistic. Actors may construct a plurality of different and incommensurate worlds.

Is the world of EDF in 1973 as real or true as any rival construction? Is the fuel cell a fiction? Are the new users who are ready to give up cars with internal combustion engines simply phantoms? Have the interests of the CGE been appropriately analysed? For almost three years everyone is uncertain. The EDF's answers to these questions go unchallenged. Even Renault remains silent, terrified by the possibility of a world that it does not have the power to unmake: a world in which it will have to undergo a difficult process of change. Is the VEL viable? This depends upon the capacity of the EDF to keep Renault in its role, prevent the contamination of catalysts, and render the new demands of consumers durable. But will Renault stay in its subsidiary role? Or will it fight back? In fact, as the story unfolds, Renault does indeed struggle with the EDF; it attempts to build its own and very different world. In short, like EDF before it, it tries what we propose to call translation (Callon, 1975, 1980b, 1985). Translation may be treated as falling into three components.

The Translator-Spokesman

EDF translates Renault, EDF translates fuel cells, and EDF translates consumers. All these expressions signify the same thing. EDF attributes to Renault an identity, interests, a role to play, a course of action to follow, and projects to carry out. EDF characterises fuel cells, the way they work, their performance, and their mode of use. The notion of translation is necessary because this enrolment (in the strict sense of occupying a role) is neither pre-given nor an external (if hidden) reality that a shrewd EDF attempts to understand and capture. Renault can be something completely different. It might be seen as a car-maker whose future depends upon the development of the petrol-engined vehicle, the automation of the assembly line and the application of electronics to the automobile. But for several years this choice does not exist. Renault becomes a company whose interests and abilities relate to the construction of car bodies for the VEL. Thus EDF translates both the will and the projects of the powerful Billancourt firm just as it translates the needs, expectations, and demands of consumers and the characteristics and performance of accumulators and fuel cells.

The translator is thus the spokesman of the entities he constitutes. EDF speaks in the name of Renault, of consumers, and of fuel cells just as a politician or a political party speaks in the name of the constituents or the social classes he aims to represent. The translator expresses their desires, their secret thoughts, their interests, their mechanisms of operation. This is the most general way of expressing it, for what is true for human entities, whether they be collective or individual, is also true for the other elements that constitute an actor-world. EDF also speaks for accumulators, for cells, for electrons, and for catalysts. By establishing the characteristics of lead accumulators, by describing the behaviour of an electron upon a catalyst, by demonstrating that a cheaper catalyst might replace platinum, by projecting the future of fuel cells and so on, EDF determines the identity of these elements and regulates their behaviour and evolution.

Translation is at first an endeavour. Later, it may be achieved. Thus Renault can block the translation and define its future differently. This, as intimated above, is exactly what occurred a few years later when it published a book arguing against EDF and its VEL. Everything is challenged by Renault: EDF's version of the interests and projects of the car firm, the possibility of constructing high performance cells, the expectations of users, and the doubtful future of the internal combustion vehicle. The actor-world of the EDF, together with the VEL, begin to fall to pieces. Meanwhile, Renault has accumulated electrochemical expertise, made contacts with the administration and obtained evaluations of the reactions of consumers. It is thus able to stop being an onlooker and to counter-attack instead. The VEL existed in 1973. In 1976, it was attacked on all sides and now exists only in the limited form of a commercial vehicle equipped with lead accumulators. Translation becomes treason, *tradutore–traditore*, once an enrolled entity refuses to enter the actor-world in order to expand into others. Since entities are not easily translated, the destiny of most spokesmen is thus to be brutally contradicted.

Translation builds an actor-world from entities. It attaches characteristics to them and establishes more or less stable relationships between them. Translation is a definition of roles, a distribution of roles and the delineation of a scenario. It speaks for others but in its own language. It is an initial definition. But, as the example of Renault shows, no translation can be taken for granted for it does not occur without resistance. Catalysts, Renault, users, and the CGE are not entities whose elasticity and 'good will' are limitless. The EDF must deal with them and test its strength against theirs. In theory, catalysts make the fuel cells work. In practice, the EDF must reckon with the fact that they may be too expensive or become contaminated (which is what scientists and economists of the Institut français du Pétrole (IFP) constantly tell EDF). In theory, Renault can be transformed and abandon the petrol-driven vehicle in order to be pulled into the actor-world of EDF. In practice, this modification proves to be negotiable only with difficulty. The same is true for the municipalities, the ministries and the users. Translation cannot always be taken for granted, and the strategies used depend upon the particular circumstances in which they develop. Certain municipalities are prepared to accept the role prepared for them and certain electrochemical generators are already available for the VEL. But Renault drags its feet, accumulates resources, and ends up challenging the EDF. Successful translation depends upon the capacity of the actor-world to define and enrol entities which might challenge these definitions and enrolments.

Translation: A Geography of Obligatory Points of Passage

If the success of translation can never be taken for granted, how is it that sometimes it is achieved? This question brings us to the second component of translation: the strategies by which an actor-world, in this case EDF, renders itself indispensable. Renault, the CGE, new users, fuel cells, and accumulators, none of these have any future outside this actor-world. The VEL thus becomes a passageway through which all the other entities that make up its world must pass. To translate, then, is to oblige an entitity to consent to detour.

This is done by choosing from a spectrum of methods that ranges from seduction to pure violence by way of simple bargaining. In the case of science and technology, one of the most generally used strategies may be called problematisation. EDF says to its users: if you want to solve your pollution and transport problems we must first create an electric vehicle. But if we are to solve the problem of how to build an electric vehicle we must first solve the problem of the short life of the electrochemical power sources that will be needed to equip the electric vehicle. Research institutes can extend the translation further: to improve the performance of electrochemical sources of power, you must first pass through our laboratories and scientific teams which are studying hydrogen catalysis on platinum surfaces. The translation thus maps out a geography of necessary points of passage for those elements who wish to continue to exist and develop. In turn, the laboratory in the above example places itself at a strategic point through which the actor-world must pass. Users of the VEL may consider that their future, their values, and their projects all depend upon the wisdom of a researcher whose eyes are fixed upon the data produced by a nuclear magnetic resonance (NMR) machine. This, in part, explains the source of the power of the laboratory in its isolation: it may become an obligatory passageway.

Translation as Displacement

While translation determines where the points of obligatory passage will be located, this does not exhaust the action. The translator-spokesman and the translator-strategist who impose certain itineraries are bringing about movement. Thus the third component of translation is displacement, in a literal sense. Some link is necessary to make entities accept certain spokesmen and certain points of passage. Otherwise, the action at a distance involved in enrolling Renault, CGE, and the fuel cell into the EDF's projects will remain mysterious.

Let us start by listing some displacements. Entities are converted into inscriptions: reports, memoranda, documents, survey results, scientific papers. These are sent out and received back, acted upon and reacted to. EDF tries to orchestrate the circulation of inscriptions, as well as the movement of people. It organises meetings, symposia, study sessions at which different parties are bodily convened. There are also movements of materials and of money. Translation cannot be effective, i.e. lead to stable constructions, if it is not anchored to such movements, to physical and social displacements (Latour, 1985, Law, 1985d).

EDF is a centre of translation, not just because it attempts to organise and structure the movements, but also because it actually ensures that it has its own centres of communication and aggregation, such as research centres, meeting halls and construction facilities, that focus and control the displacements. In this way, it extends the future of its actor-world through symposia, experiments, construction of prototypes, trial runs, investments. The actor-world accumulates materials that render it durable.

To appreciate the importance of durability, imagine that EDF had been successful and that the VEL had been finally marketed. Once built and in circulation, it would have constrained producers and consumers, central government and municipalities, within the roles created for them. The user would then have been trapped, independently of further action by EDF. He could either have accepted his fate and put himself behind the wheel of a VEL, or he could have chosen to walk.

In summary, to translate is to speak for, to be indispensable, and to displace. All translation works to solidify actor-worlds. Successful translation quickly makes us forget its history. If the VEL had been successful, the actor-world that shaped it would have lasted longer than the strategies, intrigues, power plays and trial runs that were used to build it. All that would have been left would have been producers, consumers, supply and demand, and a watchful government. The notion of translation recalls all the work and the consent that was granted, that was needed in order to achieve the seemingly natural order, where each element relates with the others.

THE CONCEPT OF THE ACTOR-NETWORK

In the preceding sections the heterogeneity of the elements that go to make up actor-worlds has been stressed, and we have considered the mechanisms by which these may be enrolled. This has made it possible to understand how both a society and the technical objects that make it up are simultaneously given shape. However, a number of additional questions have to be considered. In particular, little has been said of the structure of actor-worlds or the precise way in which they evolve. It is clear that an actor-world may be more, or less, extended, heterogeneous and complex. How shall we describe this range of possibilities, and the translations that occur between them? In order to answer this question, we introduce the notion of actor-network. This concept allows us to describe the dynamics and internal structure of actor-worlds.

Simplification is the first element necessary in the organisation of an actor-world: indeed it is an inevitable result of translation. In theory, reality is infinite. In practice, as a result of the translations that it brings about, an actor-world is limited to a series of discrete entities whose characteristics or attributes are well defined. The notion of simplification is used to account for the reduction of an infinitely complex world by means of translation.[3] For example, towns consist of more than public transport, the wish to preserve town centres and the town councils that constitute their spokesmen. They differ from one another with respect to population, history and geographical location. They conceal a hidden life whose anonymous destinies interact. However, so far as the EDF is concerned, they may be reduced to a transport system that must avoid adding to the level of pollution and a town council that seeks to advance towards this goal.

EDF does not need to know more. This definition will remain realistic so long as the simplification on which it is based is maintained. In other words, such simplifications will be maintained so long as other entities do not appear which render the world more complex by rejecting the reality represented by these simplifications as an impoverished betrayal: the town council is not representative; living conditions in different neighbourhoods cannot be reduced to those in the town centre; and the system of public transport is but one aspect of a larger urban structure. The same is true of fuel cells. If the catalysts and electrolytes that were trusted are contaminated or destabilised, the fuel cell that we hoped would power the VEL becomes appallingly complex. Instead of being easily mastered, fuel cells are transformed into an apparatus whose ever-increasing elements turn out to be beyond control. A 'black box' whose operation has been reduced to a few well-defined parameters gives way to a swarm of new actors: scientists and engineers who claim to hold the

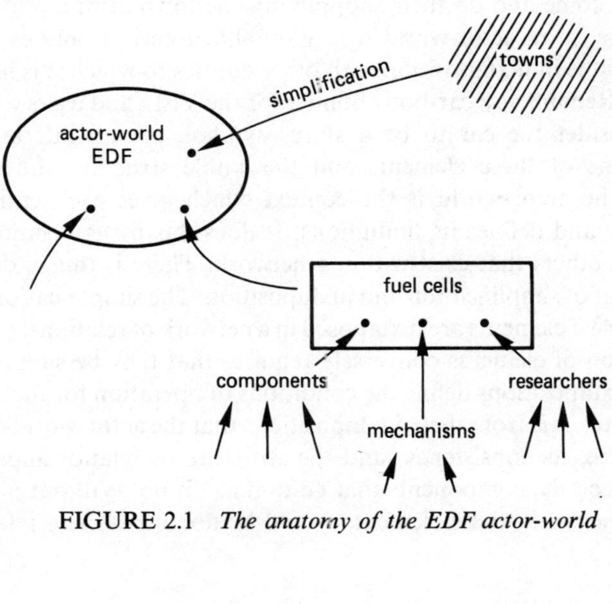

FIGURE 2.1 *The anatomy of the EDF actor-world*

key to the functioning of the fuel cell, hydrogen atoms that refuse to be trapped by the cheaper catalysts, Third World countries that raise the price of precious metals, etc.[4]

Behind each entity there hides a set of other entities which it more or less effectively draws together. We can see and know little of them before they are unmasked. Hydrogen fuel cells and zinc/air accumulators are two of the elements that make up the actor-world of EDF. However, the controversies that developed in their name rapidly divided them into a series of other elements (much as a watch is dismantled by a jeweller to find out what is wrong). Thus simplification, like translation, is never guaranteed. It must always be tested. The catalyst gives way and the fuel cell breaks down thus causing the downfall of the EDF. As for the catalysts, the electrolytes can be decomposed into a series of constituent elements: the electrons on the platinum and the migrating ions. Each of these elements will only be revealed if they are brought into a controversy: in other words, into a trial of strength in which the entity is suspected. Of course, what we have to say about fuel cells, catalysts, and the electrons is also true of city councils or administrations. In the actor-world of the EDF, the city is reduced to the city-council-that-wants-to-preserve-the-town-centre-at-all-costs. But to preserve its integrity, the city council must stabilise the elements which hold it together: the middle-class electorate that trusts it, the pedestrian precinct that pushes the flow of traffic to the edge of the town centre, the urban spread, and the system of public transport which enables the inhabitants of the suburbs to come and do their shopping in the town centre.

An entity in an actor-world (i.e. a simplified entity) only exists in context, that is in juxtaposition with other entities to which it is linked. Fuel cells, Renault as a car-body builder for the VEL, and users who no longer consider the car to be a status symbol, are all interrelated. Remove one of these elements and the whole structure shifts and changes. The actor-world is the context which gives each entity its significance and defines its limitations. It does this by associating the entity with others that exist within a network. There is thus a double process: that of simplification and juxtaposition. The simplifications are only possible if elements are juxtaposed in a network of relations; but the juxtaposition of elements conversely requires that they be simplified.

These juxtapositions define the conditions of operation for the actor-world. In fact, it is from these juxtapositions that the actor-world draws its coherence, its consistency, and the structure of relationships that exists between the components that go to make it up. Without placing them in a network, these elements would be doomed. These relationships, which define the contribution of each element as well as the solidity of the construction as a whole, are very varied. One must abandon the conventional sociological analysis that tries to adopt the easy solution of limiting relationships to a restricted range of sociological categories. Of course, there may be exchange relationships (the user exchanges his money for a VEL), subcontractual relationships (CGE works for EDF), power relationships (EDF brings Renault to its knees), or relationships of domination. But often the relationships between entities overflow simultaneously into all these categories, while some escape completely from the vocabulary of sociology or economics. How can one describe the relationships between fuel cells and the electric motor in terms other than those of electric currents or electromagnetic forces? Not only is the actor-world composed of heterogeneous elements, but their relationships are also heterogeneous. Whatever their nature, what counts is that they render a sequence of events predictable and stable. Hydrogen feeds the fuel cells which power the motor that ensures the performance of the VEL for which the users are willing to pay a certain price. Each element is part of a chain that guarantees the proper functioning of the object. It can be compared to a black box which contains a network of black boxes that depend upon one another both for their proper functioning as individuals and for the proper functioning of the whole. What would the battery be without hydrogen? What would become of the consumer without his VEL?

Therefore the operations that lead to changes in the composition and functioning of an actor-world are extremely complex. The extent to which an entity is susceptible to modification is a function of the way in which the entity in question summarises and simplifies one network on behalf of another. If we wish to construct a graphical representation of a network by using sequences of points and lines, we must view each point as a network which in turn is a series of points held in place by their own relationships. The networks lend each other their force. The simplifications which make up the actor-world are a powerful means of action because each entity summons or enlists a cascade of other entities. Fuel cells mobilise catalysts, electrons, and ions which all work for the fuel cell. This, in turn, works for the VEL and EDF actor-world. Through these successive simplifications (which are never as apparent as when they fail) electrons, specialists at Renault, the middle-class electorate and researchers at the CGE have all been enlisted, translated, and mobilised. EDF only sees and knows fuel cells, accumulators, city council spokesmen, and the public transport authorities. But each of these entities enrols a mass of silent others from which it draws its strength and credibility. Entities are strong because each entity gathers others. The strength of EDF and the durability of the VEL are built by means of these simplified and mobilised entities. Thus a network is durable not only because of the durability of the bonds between the points (whether these bonds concern interests or electrolytic forces) but also because each of its points constitutes a durable and simplified network. It is this phenomenon that explains the conditions that lead to the transformation of actor-worlds. It is possible to modify the performance of fuel cells to account for the new demands of users only if the catalysts or electron spin states can be modified in order to increase, for example, the power and longevity of the fuel cell. Each modification thus affects not only the elements of the actor-world and their relationships, but also the networks simplified by each of these elements. An actor-world is a network of simplified entities which in turn are other networks.

Transformation thus depends on testing the resistance of the different elements that constitute the actor-world. Is it easier to change the expectations of the users, the demands of the municipalities, the interests of Renault, or the longevity of platinum? This is a practical question that is answered through the continual adjustments that are also negotiated changes. To adapt the VEL by changing this or that aspect of its performance is to act upon the actor-world, and its success thus depends upon the capacity to test certain resistances to their limits whether these spring from social groups, cash flows, or electrodes to be improved.

An actor-world, such as the one described in this chapter, can in turn be simplified. The solidity of the whole results from an architecture in which every point is at the intersection of two networks: one that it simplifies and another which simplifies it. It can be translated into other actor-worlds. For example, the VEL can be linked to the TGV (high-speed train) or the Airbus, thus forming a part of a new French transport policy. Although simplified into a point and displaced in this manner, it is still composed of associated entities. While these entities are susceptible to being moulded or shaped, they in turn may transform the actor-world of which they form a part. It thus deserves to be called an actor-network. However, it is distinguished from a simple network because its elements are both heterogeneous and are mutually defined in the course of their association. Entities may disappear in order to permit the networks that they simplify to expand and surface. This capacity of self-definition and self-transformation is underlined by the fact that the two words 'actor' and 'network' are linked together in a single term. However, an actor-network is distinguished from a simple actor by its texture or structure which is an arrangement of constituent elements that has been translated. Remove all that the VEL translates and it becomes an entity without strength, society or future and it can no longer function. The actor is an association of heterogeneous elements each of which associates its own elements.

To summarise, the terms actor-world and actor-network draw attention to two different aspects of the same phenomenon. The term actor-world emphasises the way in which these worlds, built around the entities that create them, are both unified and self-sufficient. The term actor-network emphasises that they have a structure, and that this structure is susceptible to change. Accordingly, in later chapters the two are used interchangeably.

APPLICATION OF THE METHOD

In the introduction to this chapter it was asked how we might describe the 'co-evolution' of science and society in order to explain the effectiveness and the influence of the former. This chapter has put forward certain instruments of analysis which offer the elements of a response to this question.

We have suggested that the notion of translation makes it possible to describe the mechanisms by which actor-worlds are constructed. It also reveals that mechanisms cannot be taken for granted: whether they are fuel cells, catalysts, users, or industrial firms, translated entities could in theory follow other routes or be brought into other projects. They could in other words escape the logic of the actor-world into which they have been enlisted.

As we have noted, the functioning of actor-worlds and their translations are not adequately described by the usual frameworks of analysis. There are, in fact, two major obstacles that stand in the way of this: (a) the elements from Nature and Society are associated with one another in an entirely heterogeneous manner and cannot possibly be distinguished from one another (Law, 1985b); (b) the choice of entities and the way in which they are associated cannot normally be predicted because these result from operations of translation that are in progress. In short, not only does the repertoire of translated entities extend beyond that generally accepted in social science, but the composition of this repertoire does not obey any definitive rules. How can the social elements be isolated when an actor-world associates the spin of an electron directly with user satisfaction? How can any interpretation of social interaction be established when actor-worlds constantly attempt to transform the identities and sizes of actors as well as their interrelationships? The fact that actor-worlds constantly create new combinations of entities renders this task even more difficult.

The notion of actor-network is developed in order to handle these questions. This notion makes it possible to abandon the constricting framework of sociological analysis with its pre-established social categories and its rigid social/natural divide. We have shown that by inserting the entities into a cluster of heterogeneous relationships, an actor-world places them into a network. Each entity is thus reduced to a few properties which are compatible with the relationships established between the entities: Renault, a company-that-builds-car-bodies, can cooperate with CGE, a firm-that-produces-engines-and-transmissions; electrons are elementary-particles-which-transport-electric-charges-from-one-electrode-to-another thus producing the electric-current-which-drives-the-motor-of-the-VEL. An entity, firm, technical device, or social group is assigned to a black box and the actor-world may thus be seen as a bundle of black boxes. We have seen that these simplifications may always be rejected: a social group enrolled and reduced to a few interests and/or a few needs can define itself differently; a fuel cell trimmed to a few elements whose characteristics are well known can suddenly become extremely complex. Actor-worlds are supported by all the elements which have been gathered and simplified within the translated and enlisted entities. It is by virtue of this fact that they obtain their force, and if disintegration occurs, they are, accordingly, weakened.

NOTES

1. In this chapter the term battery will be used as a generic term to cover all portable chemical devices for generating electricity.
2. For the notion of enrolment, see Callon and Law (1982).
3. There is an analogy here with scientific theory. As Hesse (1974) has so persuasively argued, description always entails loss of information and simplification. The latter is thus not necessarily a Machiavellian tactic.
4. On the notion of black-boxing as a form of simplification, see Callon (1981b) and Law (1984b).

3 Laboratories and Texts

JOHN LAW

ON LABORATORIES AND RATS

What is so special about a laboratory? Why is it the laboratory rather than the boardroom that counts as Touraine's strategic locus for the twentieth century? Why is the laboratory the most important centre of translation? What goes on behind its doors that gives it the power to influence events far beyond its walls? For certainly, if you stand outside those doors, it does not seem very remarkable. The laboratory that this author knows best is up six flights of stairs in some temporary rooms that appear to have been added as an afterthought to a more permanent structure.[1] The whole is in need of a new coat of paint. At the top of those stairs you come to a double door. Until recently this door stood open during working hours. Now there are anxieties about the activities of the animal liberation front, so it is necessary to be inspected by a technician inside. You enter the hallway, pass down a corridor and walk into the main laboratory. Here there are five or six benches, and a clutter of sinks and cupboards. One or two young people are working. They are shaking test tubes, or pouring out solutions of chemicals. There is occasional talk: when are the new copolymers expected? Will the water supply be cut off before midday? Who won the game of bridge last night?

I ask for Rose. She is the scientist that I know best. She appears from a back room off the laboratory. She is wearing a white coat and latex gloves. She beckons me and I follow her into what they call the operating theatre. Here the rats are sacrificed. Rats? Well yes, they arrive at the laboratory day by day in little cages. They come from a place called the 'animal house'. The more cynical call it 'the zoo' and refer to the fact (if it is a fact) that the Home Office requires higher standards for the housing of rats than of people.

Rose is learning a new technique. She is trying to inject a radio-labelled polymer into the femoral vein of the rat. This is tricky. The femoral vein in a rat is tiny – say about the diameter of the wire that makes up a good sized paper clip. First she has to anaesthetise the animal. She does this both for her own sake and that of the rat. She does not like hurting rats but she also knows that an angry rat is not to be trifled with. So the rat is placed in a jar with some ether. First it gets excited, and then it 'goes to sleep'. She lifts it out by the tail, places it on the paper which covers the bench, and pats a pad of cotton wool containing ether round its nose. She will monitor the behaviour of the rat from now on. Too little ether and it comes round. Too much ether and it dies. It is a knife edge. She does not want to sacrifice a rat for no result.

Now she exposes the femoral vein. I look over her shoulder. She has been talking up to now, explaining what she is doing. Now she stops. Placing the point of the syringe within the vein is indeed a tricky task. She pushes the plunger of the syringe and picks up a radiation monitor. If the injection has been successful then within a few seconds there will be a count that is detectably above background in the thorax of the rat. But no, the monitor suggests that the radioactivity is all localised at the site of the injection. I keep quiet as she sighs, checks to make sure that the rat is well and truly under and prepares the other leg for injection. This time it is successful. The monitor gives an even reading right round the animal.

Now she starts to talk again. She explains that she is interested in the fate of the radioactive-labelled copolymer that she has just injected into the rat. How quickly will it be cleared from the bloodstream? And which part of the body of the rat will it go to? Perhaps it will be attracted to the tissue of a particular organ – for example the liver or the spleen. Or perhaps it will circulate in the blood, only slowly being absorbed in a uniform manner by all tissue. Now, two minutes after the successful injection, she takes a blood sample from one of the footpads of the animal. She places it in a heparinised tube. She is going, she says, to take further blood samples at five, ten, twenty, thirty, forty, fifty and sixty minutes. Then at the end of the experiment, at sixty minutes, she will sacrifice the rat and remove the lungs, the kidney, the spleen and the liver.

We sit and talk. She is quite happy to have company because she has to monitor the rat continuously to make sure that it is properly anaesthetised and she has to take blood samples every ten minutes. On the other hand, there is not a great deal to do. We discuss the rationale for the experiment. It is part of a series of experiments on a family of copolymers that has been prepared by a Danish collaborator. Today she is working on copolymer 32. She suspects that it, or some of its close neighbours, may be attracted to particular tissue types. This could be important because the copolymer in question might be a candidate for a clinical drug delivery system. Obviously no one in their right mind would ever inject radio-labelled copolymer into patients simply to see where it ended up. But the radio-label allows her to follow the trajectory of the copolymer in the body of the rat. The rat and its radioactive copolymer thus constitute what she calls a 'model system' for the human being and a drug-carrying copolymer. The idea is that if you can bind a drug to a copolymer that is differentially attracted to (say) the liver, then you can create a drug delivery system that can be targeted to that organ. The importance of this, she goes on, is that many drugs, particularly those that are used in the treatment of cancer, are unacceptably toxic to the patient at a clinically effective dose. If, however, 50 per cent or 80 per cent of the drug can be delivered to the point where it is needed, then the total dose can be reduced with no loss in clinical efficacy. She cautions, however, that there are many problems to be overcome. The copolymer itself must not, for instance, be toxic, and the cells that take up the copolymer–drug complex must be able to detach the drug from its carrier or the former will be rendered pharmacologically inactive.

I sit with Rose as she does several experiments. She sacrifices the rats, removes the chosen organs, washes these to remove blood and places them in tiny flasks. They will, she says, be homogenised in distilled water and a predetermined volume of each, together with the blood samples, will be placed in the radiation counter overnight. This is a semi-automated machine which measures the amount of radioactivity given off in a given time by samples placed in its innards. These readings are much more accurate than those of the radiation monitor. Furthermore they are printed out on paper. So long as Rose is able to keep track of the samples and not mix them up, she then has the raw data necessary to make a calculation of the body dispersion of the copolymer she has been testing. There are many correction factors to be taken into account. For instance, the organs removed contain blood which in turn contains radioactive polymer that has not been taken up by the organ in question. Nevertheless those figures, averaged with other figures from other identical experiments (she always does at least three) and compared with control results are the object of the experimental work this afternoon.

Later we have a cup of tea. Rose talks again about the copolymers she is testing. One of them looks very promising – preliminary indications suggest that it is attracted straight to the liver. This is an exciting result – it means that it may be possible to build a liver-targeted drug delivery system. But there is a further nagging thought at the back of her mind. A charitable institution called the Cancer Research Council are at the moment paying her salary and the expenses of research. They will not, she thinks, be interested in supporting too much 'basic science'. Of course, they recognise that the elucidation of the mechanisms of drug transport and targeting is important, but in the final analysis they are interested in chemotherapeutic treatments for cancer. Targeting to the liver is all very well – indeed it is an important result – but what is really needed is a system that targets differentially to cancer cells. In the present experimental arrangement with rats it is not possible to say anything specific about cancer cells. With the Council, as she puts it 'breathing down her neck' she has in mind that at some time in the not too distant future it will become necessary to conduct experiments on human cancer cells. This will, in turn, mean that it is necessary to cultivate such a cell line. This is a type of work that Rose at present knows little about. Though she wants to move in this direction, it will mean a great deal of literature research and, what is more tricky, the setting up of a new experimental system with all its attendant pitfalls. This, however, is work for the future. Today the radiation counters are at work and tomorrow there will be lists of figures on her desk.

FIRST COMMENTARY

Rose is a kind of entrepreneur. She fits together grants and people (for she has a research technician to help her in the experimental work). She also fits together animals such as rats, instruments such as scalpels, machines such as radiation monitors and chemical substances such as copolymers and radioactive isotopes. Other researchers – for instance her Danish collaborator – appear together with such bodies as the Home Office and the Cancer Research Council. Finally there are therapeutic events (clinical trials), diseases (cancer), physiological consequences (toxicity) and biochemical processes (absorption). Had the description of her activity and talk been further extended the list could have been much longer. We would, for instance, have found her discussing pinocytosis and lysozymal enzymes, the Head of the Institute, the difficulty of scheduling a meeting with the Danish collaborator, rat yolk sacs, radio-labelling and research students. The conclusion is that it is not possible to understand what is special about the scientific laboratory unless the extraordinary heterogeneity of the tasks of the scientist is recognised. Where else are such diverse elements brought together?

Where else are machines, social institutions, texts, natural phenomena such as diseases, animals, money, chemicals and people all assembled into a whole? The power of Rose, and her scientific colleagues lies in part in their capacity to call upon and juxtapose more or less any type of resource known to humankind. It thus partly lies in their extraordinary capacity to translate resources from near and far.[2]

But we may go further than this. First, if we are to understand what Rose is trying to do, we cannot think of the laboratory as a space bounded by four walls. The laboratory is simply a site where a heterogeneous range of resources from both near and far are brought together and assembled into a hopefully coherent whole. But this raises a second point. There is no guarantee that the capacity to summon up the right resources will be available. Neither is there any guarantee that they will fit together into a coherent whole. Thus we may say that the aim of the scientist is to create a mutually reinforcing structure within which one resource may be deployed to summon up another. Accordingly Rose's task, like that of the EDF, is to create and maintain such a mutually reinforcing structure. Thus it is important to inject the radioactive copolymer straight into the bloodstream. If the former escapes from the syringe into the tissue then the experiment fails. Needle and blood vessels between them do not carry the polymer to the organs that are being tested. Again, as mentioned earlier, the rat has to be maintained for an hour on a knife edge between consciousness and death. Its state of muscular tension has to be monitored and more or less ether added to the anaesthetic pad. Ether, plus rat, plus polymer, plus injection (in fact the list is endless) together channel radioactive chemicals to certain organs. If any part of this equation or any one of its constituent terms breaks down then the structure of resources that Rose is trying to put together falls apart and the experimental translation fails.

But Rose not only juggles at the laboratory bench with objects that are physically present. Again like the EDF, she also juggles with social institutions, people, results and grants. Thus the Cancer Research Council has to be persuaded to carry on paying for research. But to persuade it, it may be important to produce not only results, but results that pertain to the treatment of cancer. If this juxtaposition of the right kind of results with the right kind of social institution fails, then grants are no longer forthcoming and the whole carefully constructed edifice of resources that makes possible the production of results about the affinity of copolymers for different tissue types in the rat comes apart. The unconscious rat may balance on a knife edge, but so, in a real sense, does Rose's existence as a productive scientist.

She juggles resources. We might say that she juggles resources to create resources. There is, as it were, a kind of tree of resources creating resources creating resources. Rose is at the top of the tree. Like EDF, she is combining a cascade of high level resources to create higher level resources. As she holds the syringe to the femoral vein she does not consider the problems of breeding rats or making syringes. Or the problems of buying food for rats or maintaining the machine that presses out syringes. Or the problems of paying workers to work the machines that make the syringes and the food. It is bad enough to have to take syringes and rats and conduct satisfactory injections. There is an area about which she worries and within which she juggles resources. Then there are resources which, though themselves created as a result of similar entrepreneurial activity, she nevertheless treats as simple resources – black boxes which work in a particular way.[3] Sometimes these resources fail to perform in one way or another, and the black boxes get opened up. Thus one summer there was a strike at the British Oxygen Company which supplies bottled gas for the laboratory. Talk turned to locating other sources of gas, to industrial relations at BOC, to other ways of conducting experiments and to rescheduling work to conserve the remaining supplies. The focus of inquiry may, therefore, move down the tree of resources. Generally, however, it moves up. The aim is to juxtapose, to simplify and to routinise – in sum to build an actor-world whose parts are relatively stable.[4] After all, the injection of the femoral vein of the rat is not itself important. What is important is the study of the interaction between organ tissue and copolymer. The technique may remain a tricky one, but the idea is to perfect it to the point where it will work every time, to the point where it may be treated as a simple resource. Let us talk some more about simplification.

ON RESULTS

Rose is in her office. She has a large exercise book open at a page containing a neat array of figures. Some of these figures are from the radiation counter. They relate to the amount of radioactivity to be found in the blood, the organs, the faeces and urine, and the original undiluted sample of polymer. These she has taken from the printout and stuck into her notebook. Others, which are written in, are derived from measurements that she took when she weighed the rats at the beginning of the experiments and the individual organs at the end. Yet others represent the combined volume of homogenised organ and distilled water on the one hand, and the volume of faeces and urine on the other. She is, as she puts it, working out her results. It is a kind of book-keeping exercise undertaken with the help of a pocket calculator.

First she works on the quantity of radioactivity in the blood as this has been revealed by the counter. To do this she calculates the total volume of blood in the animal by looking at its body weight and using an empirical formula which relates the two together. Next she multiplies the count for each blood sample by the total volume of blood. This gives her a figure for the total radioactivity in the blood at the moment the sample was taken. Next she takes the radiation count for a particular organ. This she multiplies by the total volume of homogenate. This gives her a count for the organ as a whole. But the organ contains blood. Accordingly she corrects for this by noting the weight of the organ and calculating, according to an empirical formula, the volume of blood contained therein. Since she has already determined the amount of radioactivity per unit volume of blood, she is able to calculate a figure for the radioactivity of the blood contained within the organ and deduct this from the total figure of the organ to generate a final estimate of radioactivity in the organ in question. She does this in turn for each of the organs. Then, by analogous means, she determines the quantity of radioactivity in the faeces and urine at sixty minutes.

She has decided to express the levels of radioactivity not in absolute terms – there are too many imponderables and assumptions to make this a warrantable exercise – but in relative terms. She accordingly reworks her figures to express them as percentages of the total amount of radioactivity recovered from the organs and the blood at sixty minutes. To do this all the figures are added together and each is expressed over the total thereby derived and multiplied by 100. In addition the total is combined with the figure for the combined urine and faeces which is in turn divided by the figure for the undiluted sample and multiplied by 100. In this way it is also possible to calculate a figure for the total proportion of radioactive material recovered at the end of the experiment.

The end result of this process is a table that looks like this:

Relative body distribution of radioactivity recovered at 60 mins

Liver	Lung	Spleen	Kidney	Blood	Recovery (%)
6.62	2.00	0.40	44.00	46.97	60.92

SECOND COMMENTARY

Juxtapose, simplify, routinise. These are the related aims of the entrepreneur in the laboratory. We saw Rose at the bench fitting objects and processes together in a tricky combination. We saw her sometimes succeeding, sometimes failing. But now we have moved with her to her office we uncover one of the secret weapons of science – the capacity to convert the diversity of living bodies, chemicals, procedures and machines into a set of figures written or printed on sheets of paper.[5] What a coup this is! But what an effort too! Buildings are erected, benches made, courses taken, examinations sat, degrees awarded, grants offered, rats bred, radiochemicals synthesised, organs chopped up and blended with water. Complexes of radioactive iodide are pumped into blood, veins pierced by needles, blood extracted at the behest of chronometers, tubes placed in machines, and ionizing radiations forced to pass through detectors. So many different kinds of bits and pieces are strung together. But the end result is a set of traces, for organs and rats never make a personal appearance in Rose's office. Neither do syringes nor femoral veins. We have moved forward on a breathtaking wave of simplification to a point where juxtaposition takes place between docile figures by means of procedures that are operated by people and calculating machines. Docile? Well not really. The figures which Rose works upon are not empty. They are equivalent to endless operations and resources and cannot easily be ignored. They are, indeed, networks that have been reduced to points in the actor-network that she is building. Naturally then, they can be awkward and intractible and it takes effort to ignore them, but at the same time if they can be successfully simplified then they add their weight to her actor-world. And if this is managed, then some of the complexity of heterogeneity disappears. These figures are simplified; all of a kind. They may be added and subtracted or forced through standard procedures. There is, of course, still work to be done, but compared with the juggling that takes place at the laboratory bench to create them in the first place their subsequent juxtaposition and conversion into a set of results is, shall we say, a task that has been simplified from three to two dimensions.

At root the role of the scientist has not changed. He or she is still translating, juggling and simplifying. But the *technology* of that translation has altered. The technology of pens and papers, of pocket calculators and computer printouts, allows a huge range of juxtaposed pieces and processes to be laid out simultaneously on the space the size of a desk. It allows their easy manipulation. And it allows their easy comparison. For now we begin to see the first comparisons. Already the behaviour of one liver may be compared with that of one spleen, a pair of lungs, a pair of kidneys and the blood in the presence of copolymer 31. Already it is possible to say something about their relative behaviour. But this achievement, important as it is, is only the palest foreshadow of what is to come, for Rose is going to build an actor-world in which everything from rats to social institutions is compared. And to do this she will make indispensable use of the simplifying technology of traces.

ON TABLES AND RESULTS

We are sitting in Jack's office. He is the Head of the Institute. Rose is telling him about the progress of the work. Once again the desk is covered with sheets of paper. Some of these contain tables and others graphs. Rose is talking. She is telling Jack that the graphs and tables refer to two different kinds of results. The first concerns 'blood clearance' – that is the decline in the level of radioactive copolymer in the bloodstream of rats over a period of sixty minutes after injection. The second is to do with 'body distribution' – that is the location of that radioactivity in different body organs and blood at the end of the sixty-minute period. Though the two sets of results relate to the same experiments, they are expressed in ways that make direct comparison of the figures impossible. Accordingly, general but not specific contrasts may be made.

She points first to the blood clearance results (see Figure 3.1). Note, she says, the way in which copolymer 31 disappears rapidly from the bloodstream. After one hour only 10 per cent of it is still circulating. It disappears, she suggests, much more quickly than the other copolymers – numbers 29, 30 and 32. Number 31 is thus an interesting case. Furthermore, this result is consistent with the body distribution data (see Figure 3.2). These suggest that it has great affinity for the liver. The two scientists study the traces but their excitement is obvious. They are delighted that something out of the ordinary has occurred. They talk excitedly of it as a 'fantastic' result and agree that it stands out in relation to comparable findings from both within the laboratory and elsewhere. Thus 65 per cent of copolymer 31 goes into the liver within an hour. In the case of liposomes, according to work undertaken by a man called Pierce from another institute, the comparable figure is only about 40 per cent. These are results that must be published – they would certainly form the backbone of a good paper.

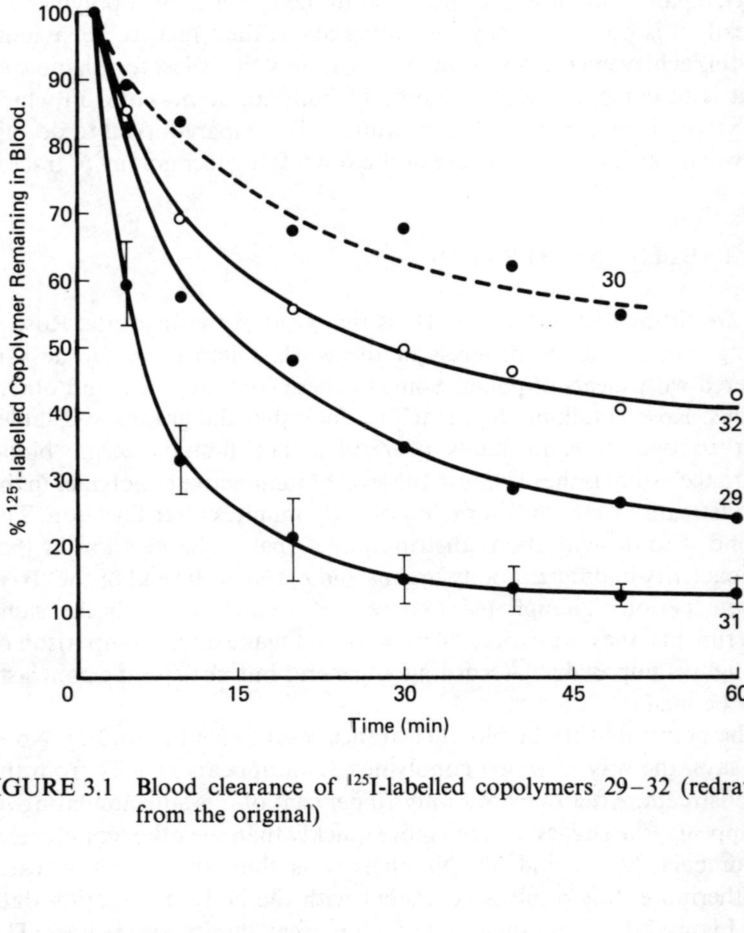

FIGURE 3.1 Blood clearance of ^{125}I-labelled copolymers 29–32 (redrawn from the original)

Not all the findings are, however, quite as satisfactory. There appears, says Rose, to be an inconsistency in the figures for copolymer 29. While the blood clearance data suggest that this is quickly removed from the blood, the body distribution data suggest the contrary. Though, as she has already noted, the figures are not *directly* comparable, on the face of it there nevertheless seems to be a troublesome inconsistency. Perhaps this may be partially explained by the low recovery rate of copolymer 29, but even so, she concludes, further work will be necessary to sort this out.

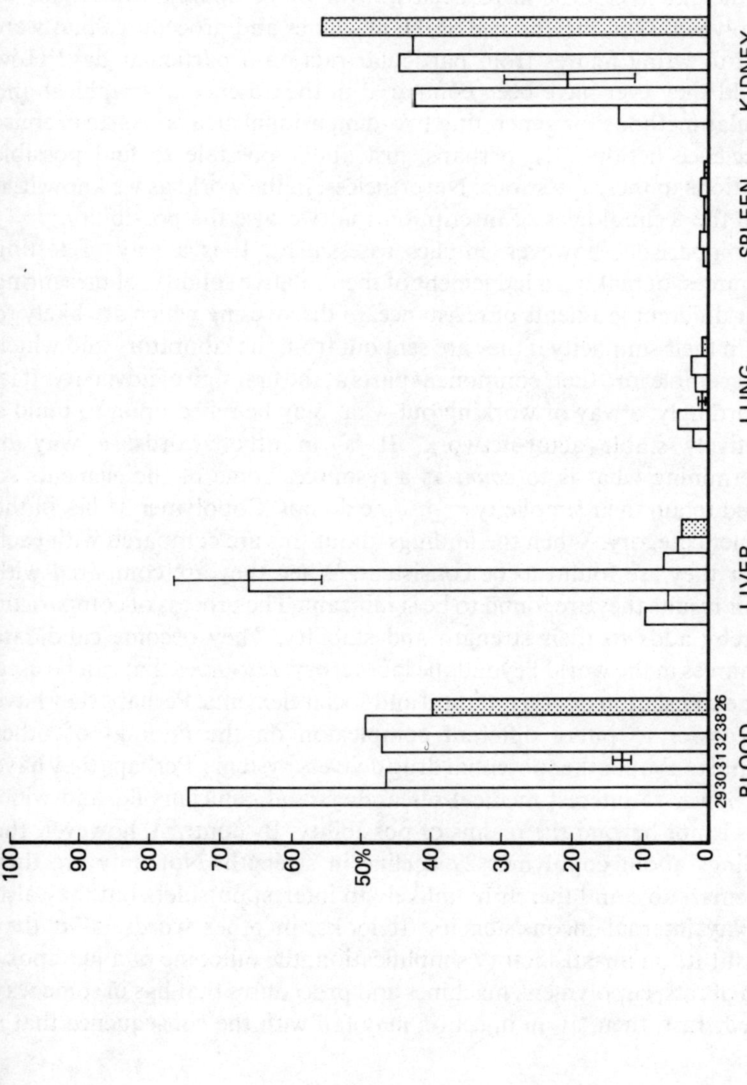

FIGURE 3.2 Body distribution of radioactivity sixty minutes after intravenous injection (redrawn from the original)

THIRD COMMENTARY

Juxtapose, simplify, routinise and compare. These processes are common to all translation. However, the technology of two-dimensional translation renders easy the juxtaposition of elements that would otherwise never come together. It is the product of a form of simplicity enforced on the heterogeneous which permits otherwise improbable comparisons and assessments. How would the effects of copolymer 31 on the rat liver ever have been known or compared with those of copolymer 32 in the absence of the machines and procedures that were used to wring figures from particular rats on a particular day? How would they ever have been compared in the absence of graphical and tabular methods for generating two-dimensional arrays? As an exercise in science fiction it is, perhaps, just about possible to find possible solutions to these questions. Nevertheless, in the world as we know it, it is to the technologies of inscription that we owe the possibility.

Comparison, however, implies assessment. It is a way of testing resources, of making a judgement of their relative solidity, of measuring their different gradients of resistance, of discovering which are likely to retain their simplicity if they are sent out from the laboratory and which will crumble into their component parts at the first sign of adversity. It is, accordingly, a way of working out what may be relied upon to build a relatively stable actor-network. It is, in other words, a way of determining what is to *count* as a resource. Some of the elements so tested retain their simplicity and some do not. Copolymer 31 lies in the former category. When the findings about this are compared with each other they are found to be consistent. When they are compared with other results they are found to be significant. The process of comparison thereby adds to their strength and stability. They become candidate resources in the world beyond the laboratory, resources that can be used in the translation of other natural and social elements. Perhaps they have the power to put a different complexion on the findings of other scientists about other potential drug delivery systems. Perhaps they have the power to interest medical researchers and clinicians far and wide. This is not beyond the realms of possibility. By contrast, however, the findings about copolymer 29 decline in strength. Not only are they unremarkable and therefore unlikely to interest outsiders but they also display internal inconsistencies. It looks, in other words, as if they constitute an unsatisfactory simplification, the outcome of a juxtaposition of rats, copolymers, machines and procedures that has in some way failed. Just, then, as an injection may fail with the consequence that a procedure may not be used as a simple resource, or a workforce may go on strike and the provision of bottled gas becomes a complex problem, so too a set of figures, juxtaposed with their cousins in a two-dimensional space, may evidence their weakness and either be abandoned or force an investigation about how this might be rectified.

ON TRANSMISSION

Rose's office is filled with paper. There are letters to and from editors. There are journals. There are reprints. There are invitations to address conferences in different places. There are early handwritten drafts of papers. There are later drafts on computer printout. There are proofs to be corrected. There are catalogues from the makers of scientific instruments. There are departmental memos to do with seminars and graduate students. There are notes from travel agents. There are shelves of laboratory notebooks. There are posters in preparation for meetings. There are slides. There are files of correspondence with grant-giving agencies. There are grant applications. There are polite refusals. Sometimes, indeed, it seems that Rose might vanish under the weight of paper in her office. It is fortunate that she is very well organised. Everything is carefully filed, she can lay her hands on anything at a moment's notice and she never gets lost in the maze of paper.

As usual she will be attending one or two conferences in the not too distant future. The posters are for one of these where she will not actually be giving a talk. But uppermost in her mind at the moment is a paper that she has to give in a month's time to a meeting in America. She is pleased to have been invited to this. It is a small, select gathering and she feels that she owes her invitation in part at least to the fact that a talk she gave to another international meeting early in the year was well received. She has been working on this important presentation for a fortnight, on and off, and is now concerned that the slides be prepared in plenty of time. However, this is not the only writing that she has to do. She has a list of papers in the front of one of her folders. Some of these are only ideas at the back of her mind. Others, however, have taken more definite form. She knows what material they will contain and has decided which journal she will try to place them in. Yet others have been written in one draft or another, or have already been submitted. Some are being revised while some are far down the pipeline: the page proofs have been sent by the journal concerned.

Today she is particularly pleased that a paper submitted to *Fun-*

damenta Polymer has been accepted. She did not expect too much difficulty in placing this for it contains the results of the experiments undertaken on copolymer 31. Nevertheless she is delighted. This successful work will now see the light of day in a journal which is specifically concerned with the effects of polymers on biological systems. Indeed, she, together with her co-authors (for both Jack's name and that of Anderson, the Dane, also appear) have written about their discovery in a way that underlines its importance for drug delivery systems. They have stressed that this finding, valuable though it is, is only part of a longer term programme to build, not model systems in which polymers carry radioactive markers, but pharmacologically active complexes that may ultimately be tested in a clinical context. Thus, in their conclusion, the authors have written that they are already testing certain copolymers which have been bound to drugs that are known to inhibit cell division.

FOURTH COMMENTARY

Cascades of heterogeneous elements are juxtaposed to create something simple and solid – a point in the network that makes up an actor-world. This is in turn juxtaposed with its neighbours to create another actor-world. The process is endless. Anyone wishing to translate, to build an actor-network, has to start somewhere, to take the solidity of some kinds of components for granted in order to build others. We saw this for the EDF, but it is equally true here. In Rose's laboratory individual experiments, machines or organisms may fail. But in general they reveal their solidity as they stand up, in juxtaposition with others, to the task they have been allocated.

I have suggested that the resources juxtaposed in the translations of science are particularly heterogeneous – that this is one of the sources of the power of science. It is capable of calling equally on animals and social institutions, natural phenomena and people and using them in its search to build new sources of power, new resources. I have also argued that its secret weapon is its capacity if not to homogenise this heterogeneity, then at least to reduce it to two dimensions by means of the technology of inscription. The laboratory may, in other words, be seen as a device for the enforced extraction of homologous and commensurable traces out of diversity. But there is a final link in this chain that has still to be forged. The specificity of the laboratory lies finally in its capacity to send out messengers that have the power to transform the world. In other words, it is not good enough that the laboratory can call upon heterogeneity as its raw material. It is not even good enough that it can homogenise this heterogeneity. It must also somehow acquire the capacity to act at a distance upon the world in all its diversity.[6]

What can the laboratory send out that will withstand the battering of the world? What kinds of juxtaposed sets of bits and pieces will not come apart once they depart? The list is very limited. People may go out. Rose may go to meetings of her colleagues, visit grant-giving bodies, and talk to voluntary associations. Indeed she might leave the laboratory and take up a post in industry. Chemicals are similarly transportable. It happens that Rose never sends these out herself, but she receives them every month both from suppliers and her Danish collaborator. Money also comes and goes and talks with authority. Firms send machines and chemicals at its behest. There is also a traffic in instruments and organisms. But, though all of these are essential resources in one way or another, the most important is the text. It is the text most of all that the laboratory uses to rebuild the world. It is the text that above all else forces others, willingly or not, to consider the importance of drug-carrying polymers in the treatment of cancer. It is the text that boxes in and regulates the points of contact between clinicians and researchers, patients and fund-raisers, polymers and livers, laboratories and diseases. In short it is first and foremost the text which imposes a structure on the world.

It is true that texts rarely go out alone.[7] They depart accompanied by people, by money, by chemical entities or by machines. A combination of two or more of these can constitute an array with a formidable capacity to translate. Nevertheless, the text is the key weapon in this arsenal. First, it is durable and extremely transportable. Second, it is reproducible and thus highly diffusible. It may act in many places simultaneously. But third, and this is perhaps the most important point, *it displays its structure in a complex 'synchronic translation' of the endlessly heterogeneous*. There they all lie, the bits and pieces, in the same place at the same time, linked together on a sheet of paper. The strength of the translation lies in what is selected and how the linkages are built. But there are no limits to what may be chosen and how this may be juxtaposed, for the power of the text is a function of the fact that it displays and arrays whatever resources may contribute to its strength. Results, beaten out of rats, machines and pocket calculators, are juxtaposed with chemical or physical phenomena, diseases, drugs, clinical trials, scientists, research foundations, chemical formulae, suppliers, the findings of different laboratories, the names of other scientists and much more besides. To act at a distance on (for instance) clinicians, the text thus assembles an array of resources that have the power to enforce a *whole world* within which the clinician is located. In other words, by their very nature texts reveal the complex structure of the actor-network as it seeks to translate others and thereby build a world.

Take the heterogeneous. Homogenise it to two dimensions. Juxtapose this homogeneity in a forceful textual structure. Send the structure out to act on others who will be likewise homogenised. This is the secret of the laboratory as an agent of change, a secret that involves building complex but enforceable worlds on paper. But the analyst who wishes to understand the struggles between actor-networks finds him/herself in a very privileged position. The structure of the worlds so built on paper are displayed for all to see. People may lie to us. Machines do not talk. But texts, once inscribed and diffused, cannot be changed. They may be rejected. They are, of course, also multi-interpretable. But for all that an essential part of their structure remains fixed and on display for all to see. How, then, does this ultimate textual weapon work? This is the subject of the next three chapters.

NOTES

1. For further ethnographic details see Williams and Law (1980); Law and Williams (1981, 1982); Law (1983); and Chapter 5 in this volume.
2. On the concept of translation, see Chapter 2 and also Callon (1980b).
3. On black boxes see Callon (1981b) and Law (1984b).
4. For another view of the importance of simplification in the laboratory see Star (1983).
5. For a seminal analysis of the role of inscription in science see Latour and Woolgar (1979). For a general analysis of the importance of texts and printing for social control see Latour (1985). For a study of the way in which texts, people and devices interact to generate power in science see Law (1985a).
6. On the capacity of the laboratory to act upon the world see also Latour (1983, 1984); Law (1984b).
7. See Law (1985d) for this point, developed in the entirely different context of Renaissance technology. See also Latour (1985).

4 Writing Science – Fact and Fiction:

The Analysis of the Process of Reality Construction Through the Application of Socio-Semiotic Methods to Scientific Texts[1]

BRUNO LATOUR and FRANÇOISE BASTIDE

Over the last fifteen years our conception of the scientific article has greatly advanced with the application of methods borrowed from history, literary criticism, rhetoric, semiotics, and finally, the microsociology of science and technology.[2] During the last few years, scientific discourse which was formerly thought to be inaccessible to laymen or written 'without literary style', has become an almost routine subject for literary criticism. In this chapter we demonstrate some of the results of these studies experimentally.

Our aim is to highlight some of the methods used by scientists to create a forceful textual structure, one that will carry weight when those who wrote it are no longer physically present to argue their case. Our chosen method is to alter a few texts in such a way that the reader can feel the different effects provoked by these modifications. We then briefly comment on the significance of these effects.

EFFECT 1: DECIDING ON HISTORY AND EPISTEMOLOGY

'Evidence for a process of counter-current water exchange in the inner regions of the hamster kidney'
by F. F. François, R. Maxime and C. Claude
Biology Department, Atomic Energy Commission. Received 22 March 1960.

[1] Over the last few years, our conception of the mechanisms of urine concentration in the kidney has greatly advanced following experimental observations by Wirz, Hargitay and Kuhn (1951) and their interpretation.

[2] In spite of the experimental observations of Wirz, Hargitay and Kuhn (1951), our understanding of the mechanisms of urine concentration by the kidney has evolved little over the last few years. Recently, however, a new interpretation of their results has sparked off decisive advances in our knowledge on the subject. This interpretation has brought to light the process of counter-current water exchange.

In paragraph [1], the flow of events is suddenly interrupted by the appearance of 'greatly advanced' conceptions. The cause of these advances is attributed to the works of three people, 'Wirz *et al.*' Finally, as is usually the case in classic epistemology, a distance is established between 'observation' and 'interpretation'. In paragraph [2] this same distance is maintained but is used to deny that 'Wirz *et al.*' are responsible for the 'greatly advanced conceptions'. The 'observations' had no value before a new interpretation explained and made sense of them. The rhythm is altered for it is the authors of the article, not Wirz *et al.*, who instigate the change.

Thus in three lines an article already expresses a history of science and an epistemology. It also distributes responsibility for this history. To go from form [1] to form [2] obviously requires lengthy discussion and arduous negotiations between the authors, and then between the authors and the referees. These negotiations are marked by erasures, interpolations and additions in the successive rough drafts (Knorr-Cetina, 1981; Williams and Law, 1980; Lynch, 1982).

EFFECT 2: SETTING UP A NEW SCENARIO

[3] The procession slowly progressed through the winding streets of the old city. From high up in the belfry, I was easily able to distinguish the little scouts, the musicians of the Sons of France, and the men from the church council carrying the canopy. The crowd was lined along each pavement and although most were only Sunday believers, they listened quietly as the Daughters of Mary passed praying. However I noticed that at every street corner the scouts, who were

impatient to get a bite to eat at the chaplain's headquarters, threaded their way with difficulty through the loiterers. They passed from one street to the next, shortcutting the procession, and then dispersed towards the fun fair. Moving from the main streets to the alleys, the procession lost its children and, little by little, was whittled down to a core of pious but middle-aged souls.

[4] The results detailed can be perfectly explained if one accepts the hypothesis that the walls of the vascular and urinary hairpins are much more permeable to water than to sodium. This would generate a counter-current water exchange between the ascending and descending limbs. If the walls of the ducts are more permeable to water, 'transversal diffusion' should cause a fraction of marked water molecules circulating in the descending limbs to pass into the ascending ones by exchange at each level.

These two paragraphs construct an observer, one in his belfry and the other in his laboratory. These observers are assumed to have witnessed a phenomenon which has the same form:

hamster kidney	old city
counter-current hairpins	winding alleys
duct walls	assembled crowd

A mixed set of elements circulates in this general decor: on the one hand, marked and unmarked water and sodium, and on the other, a procession. In both cases the crowd becomes differentiated and concentrated into either sodium at the turn of the hairpins or pious middle-aged souls at the end of the day. In both cases the sodium and the Daughters of Mary remain true to the course whereas the water and the little boys pass respectively through the walls and the crowd and move off.

Neither paragraph is more concrete, technical or simple than the other. 'Sons of France' is a local term known only to the natives of Beaune. As for the 'church council', it is certainly no better known than 'sodium'. Though few people have dissected the ducts of a kidney or observed the functioning of a system of counter-current water exchange in urine, how many Frenchmen have seen a Corpus Christi procession from a belfry? As for the separation between religious fervour and the desire to get a snack, no more is said of this than about the wall that holds sodium in but lets water through.

In both paragraphs a certain degree of concentration must be revealed to the observer and then to the reader. The first observer climbs up into his belfry and tells us what he sees, while the second descends by stages into the kidney and reports what he has observed.

At this point, though, the two observers take totally different routes. To see the layout and the flow of the old city, the first observer takes advantage of Corpus Christi. He sees the procession take place directly as a continuous event unfolding in time and space. The second observer sees nothing. The kidney is too obscure. He must create the event, build time and space, and reconstitute the actual continuous movement from discontinuous observations.

To create the event, the observer must introduce radioactive water and sodium where unmarked water and sodium usually flow. The kidney will not know the difference but the observer will. Just as with the brown shorts of the Boy Scouts and blue uniforms of the Daughters of Mary, the radioactive water and sodium mark a contrast. The relative concentrations of elements thus become visible. Of course, we still do not see anything but at least we can subtract one quantity from another.

This is not enough to construct a narrative. The person who wants to observe something happening in the kidney must create the movement of his invisible procession by superimposing photo finishes. Just as those who organise car rallies place tallymen at predetermined points along the course, the observer systematically slices the kidney into eight segments from the cortex to the papilla. Since it is impossible to follow the flow through a kidney cut into sections, the observer must start the procession in several kidneys and then interrupt the flow each time after an increasingly long interval. The still-life story of displacement is then recounted on a table.

EFFECT 3: SETTING UP INSCRIPTION DEVICES

The observer in the belfry of paragraph [3] saw the little brown spots passing through the crowd but the reader only saw a text saying that the author had seen them. Unlike the first observer, the second refers his reader to a figure (Figure 4.1) which is explained by a legend which then refers the reader to a table (Table 4.1). This double entry table is a record of what happened in the laboratory. It is obviously much more difficult to bring the counter-current to light. What do we actually see in Figure 4.1? Literally speaking, we see a comic strip. That is to say, time passes from left to right. This is an accepted convention that is easily adopted

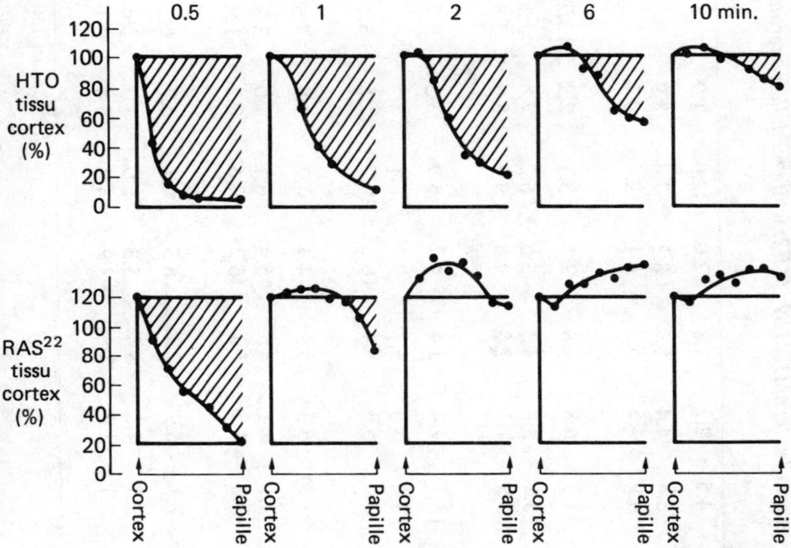

FIGURE 4.1 Turnover, with time, of water and sodium in the different regions of the kidney (redrawn from the original).
Abscissa: the different regions of the kidney as defined in Table 4.1.
Ordinates: top: the radioactivity of tissue water (i.p.m./mg) expressed as a percentage of that in the cortex; bottom: the specific radioactivity of sodium (i.p.m./μg Na) expressed as a percentage of that in the cortex. The numbers at the top of the figure indicate for each curve the time interval (min) between the injection of the isotopes and the removal of the kidney.

when reading 'Peanuts'. The event to be described in this comic strip is a decrease in the hatched surface limited by a curve. 'Everyone' is able to see that the surface diminshes from left to right. As any 10-year-old can tell, every drawing in this comic strip is also coded according to its abscissa and ordinates. The abscissa goes from the surface to the inner regions of the kidney or, for the anatomists, from the cortex to the papilla. For the ordinate, there is a scale going from 0 to 120 which would pose equally few problems since everyone knows how to read a thermometer at least.

Using such simple reading conventions, the drawings nevertheless become fairly difficult to understand. 'Everyone' can read each element but the sum of them all is easily legible to perhaps ten people. The two strips are not to be read as a continuous sequence but in contrast to one

TABLE 4.1 *C Cortex. J Cortico-medullar junction. MR1, MR2, MR3 red medulla (external). MB1, MB2: white medulla (internal) BP: Base papilla. Pl: Plasma.*

Expérience du 17 juillet 1959	HTO i.p.m./mg tissu frais × 10^{-2}	31.5	13.2	4.5	2.2	1.8	2.0	2.0	1.9	
Hamster (96 g)	^{23}Na μg/mg tissu frais	1.4	2.6	3.4	4.4	5.4	8.3	7.4	8.9	
temps écoulé 0.5 min	^{22}Na i.p.m./mg tissu frais × 10^{-2}	3.9	5.1	4.9	4.5	5.2	6.1	2.4	0.7	
Expérience du 31 juillet 1959	HTO i.p.m./mg tissu frais × 10^{-2}	17.1	16.9	11.3	6.8	4.8	2.5	3.1	2.2	12.2
Hamster (108 g)	^{23}Na μg/mg tissu frais	1.7	2.3	3.1	3.8	4.3	5.6	6.2	7.8	3.1
Temps écoulé 1 min	^{22}Na i.p.m./mg tissu frais × 10^{-2}	2.7	4.2	5.9	7.1	7.5	9.6	9.6	8.9	3.8
Expérience du 10 juillet 1959	HTO i.p.m./mg tissu frais × 10^{-2}	15.8	16.2	13.1	9.4	5.4	4.7	3.9	3.3	9.7
Hamster (110 g)	^{23}Na μg/mg tissu frais	1.7	2.7	3.4	4.6	5.7	7.9	8.1	3.7	
Temps écoulé 2 min	^{22}Na i.p.m./mg tissu frais × 10^{-2}	2.1	3.6	5.0	6.3	8.4	10.5	10.7	11.5	3.1
Expérience du 16 juillet 1959	HTO i.p.m./mg tissu frais × 10^{-2}	6.8	7.0	7.3	6.3	5.9	4.4	4.2	3.9	7.6
Hamster (96 g)	^{23}Na μg/mg tissu frais	1.4	2.8	3.1	4.1	4.8	5.9	6.7	7.0	3.7
Temps écoulé 6 min	^{22}Na i.p.m./mg tissu frais × 10^{-2}	1.3	2.3	3.2	4.2	5.3	6.2	7.6	8.2	1.9
Expérience du 31 juillet 1959	HTO i.p.m./mg tissu frais × 10^{-2}	7.1	7.1	7.7	6.9	7.2	6.5	6.1	5.7	6.2
Hamster (120 g)	^{23}Na μg/mg tissu frais	1.7	2.7	3.3	4.1	5.4	5.3	6.0	7.7	3.4
Temps écoulé 10 min	^{22}Na i.p.m./mg tissu frais × 10^{-2}	1.1	1.6	2.3	2.9	3.6	3.9	4.4	5.4	2.1

another. The upper strip describes the passage of marked water, the lower of marked sodium. To get used to reading this figure one must, as in gymnastics, undertake a series of simple movements. Each image is first read as a curve according to its abscissa and ordinate. Then the five images are read as a time sequence (towards the right). Each strip is then read in comparison with the other (from top to bottom and bottom to top). Finally, each point on the curve is transformed into the positions of the elements that travel through the hamster's kidney. This figure is neither simple nor dificult. Nor is it easy or difficult to read. According to the legend, if we agree to read it, the figure is either as 'plain as day' or 'means nothing at all'.

The legend, as the name indicates, tells us what we should see in this story of procession and sacrifice:

(a) The area diminishes from left to right and diminishes more rapidly at the top than at the bottom.
(b) In the inner regions of the kidney, the water is renewed less quickly than the sodium.
(c) Thus we see a delay. The water takes a shortcut.
(d) A counter-current is revealed.

We can question what the author claims to have shown in two ways. We can either accept the figure but refuse to go to a level higher and refer to the sense of the legend, or we can ignore the figure and go to a lower level by referring directly to the experiment.

For the first case, the reader can say that he sees the areas diminish but cannot see any 'delay', even less a 'counter-current'. He can thus say that the figure 'proves nothing'. He therefore challenges the series of transformations that allow one to pass unequivocally from the comic strip to the process of counter-current water exchange.

In the second case, the reader can challenge the figure at a lower level and deny that the points of each graph correspond to anything. That is why the authors refer the (potential) reader to the table.

The table makes use of the same visual oppositions as the figure but in a different order. The top and bottom now correspond to a reclassification of the experiments in relation to the time interval between injection and removal. Left and right now correspond to a display of the kidney slice in relation to anatomical reference points. To make sense out of the table, one must undertake a series of subtractions. It contains many more details but shows less than the figure. Nonetheless, the figure is completely dependent upon the table. It cuts out one column from the table and reduces the quantity of numbers. As a result of a series of transformations, certain ratios are established that lead to percentages. The authors therefore transform the discontinuous subtractions into a continuous reading of surfaces. Thus the reader sees things in the figure that he does not see in the table. However, if the reader is suspicious of the figure, he has to trust the table. The table consists of indices that refer to an outside world (the laboratory) where certain events seem to have actually occurred: 'Expérience du 17 juillet 1959; Hamster (96 g); Temps écoulé: 0.5 min'. The empty boxes in the columns indicate the presence of errors and uncertainties. The table is full of little details that make it look real. Unlike the illustrations that accompany normal narratives, the figure is not based on an 'external' source but only on a table which in turn confusedly points towards the world of the laboratory. Reaching this point, the reader still sees no more than when he read the narrative of the observer in the belfry. He still simply has to believe the author.

EFFECT 4: NEGOTIATING THE TEXT

[5] Figure 4.1 shows that tritiated water and radio-sodium behave differently within the inner regions of the kidney. The surface of the hatched zones for each time period is in proportion to the quantities of water and sodium that have yet to reach equilibrium with the cortex. This surface diminishes with time but follows a rather different path for water than for sodium. Indeed, ten minutes after the injection of the indicators, the concentration of tritiated water in the innermost regions is still less than that in the cortex. This indicates that in the inner regions of the kidney the renewal of water is much slower than for sodium.

The written text of the article is a verbal commentary on the figure, just as Figure 4.1 is a visual commentary on the table. After having seen the table and read the legend we should now be able to understand the written explanation. Paragraph [5] aligns a series of equivalents so that one is able to go from simple drawings to the hamster kidney. In seeing Figure 4.1, it is as if we were seeing the functioning kidney in the hamster. According to the text, the hatched zone is/represents/figures/is worth the same as the 'quantities of water'. The 'rather different' paths are/represent/figure/worth the same as 'the delay' which in turn 'reveals' the counter-current in exchange.

The text of the article does not say anything that it fails to show. It does not show anything that is not supported by the data. Nevertheless, it does not actually show anything. It builds a chain of transformations that makes it possible for us to tie the discovery of a counter-current to a hamster in oliguria sacrificed at the other end of a chain on 17 July 1959 at the Biology Department of the Atomic Energy Commission. The text shows everything and the text shows nothing. What then does it show? It presents traces, inscriptions, and prints in which lots of things are said that are summarised in one sentence by the title. Does a scientific text tell a story like other kinds of texts? No, because it stacks the traces in such a way that each one is a transformation of the last. So does it therefore constantly repeat itself since it talks about the same hamsters ten times in a row? No, because it adds something each time which is not really shown but then is not really unfounded either. The text puts together its elements as if they were stones in a primitive arch. Each one is supported by the last but leans out into the void. Yes, it is a construction. Is it fragile? Sturdy? That depends upon the masons, the thrust it must sustain, and the fit. But mostly it depends upon the negotiations which guide the way in which each stone is balanced upon the last. Too much timidity and the article will never be finished. Too much audacity and it will crumble.

EFFECT 5: FORECLOSING OBJECTIONS

[6] Scientific objector no. 1: 'If the kidneys are not immediately frozen the isotopes will diffuse everywhere and the measurements will be meaningless.'

Materials and Methods, line 9: 'The excised kidneys are *immediately* immersed in liquid nitrogen.'

Scientific objector no. 2: 'But that's not good enough! If the razor used to cut the frozen kidneys is warm, you will disturb the results!'

Materials and Methods, line 12: 'Each kidney is then dissected in a cold room (– 5°C) to avoid any thawing. A segment spanning the whole of the kidney from the papilla to the cortex is prepared using a chilled razor blade.'

Scientific objector no. 3: 'In any case, you'll have to thaw the kidneys sooner or later to prepare your samples and then the radioactive vapour will evaporate.'

Materials and Methods, line 19: 'These frozen fragments are placed in pyrex tubes that are immediately capped with an airtight stopper made of aluminium foil.'

Scientific objector no. 4: 'Yeah, but they'll have to be uncapped so that whatever solvent you want to add can be introduced.'
Materials and Methods, line 25: 'The tubes are weighed and then 1 ml of distilled water is injected through the aluminium foil with a precision syringe. With this precaution it is possible to avoid having to uncap the test tubes and thus lose highly radioactive water vapour originating from the labelled water in the tissue that has been removed.'
Scientific objector no. 5: 'You don't know what you're doing! You haven't even ground up the tissues! How can you be sure that the labelled sodium and water will be completely equilibrated within the solvent, huh?'
Materials and Methods, line 25: 'The small size of the fragments (2–10 mg) allows labelled water and tissular electrolytes to diffuse quite rapidly in the supernatant where they can be measured.'

Paragraph [6] introduces a fictitious dialogue between either the author and his 'dear colleagues' or between him and his superego(s). We have moved, therefore, from the text to a dispute about where the text comes from. It is here that we reach a weak point, the Achilles' heel, of the whole article. This Colossus may have feet of clay, this counter-current may flow through a paper kidney. Thus the table is supported by another section which is written in prose and in small print. This section has long been known as Materials and Methods. It is the most polemical section of the entire article, but it is a polemic that is so intense that it ends up looking very boring.

Imagine that, in order to convince us, the authors had only presented the numbers in Table 4.1. After all, that should have been good enough. These authors seem to have been honest and we should have been able to trust them. However, had this happened, we would have immediately heard an outcry of indignation and protest. Where do your numbers come from? What is their genealogy? How did you remove the kidneys? What was the temperature of the razor? Were the tubes tightly capped? Each of these colleagues' questions points at a gesture to be made, a precaution to take, an instrument to buy and a skill that is needed. All of these are absolutely necessary if one wants to be able to lead a hamster kidney to reveal its functioning. It is as if someone were making a long water hose by connecting several shorter ones together. One knot, one defective joint, one leak and not a single drop of water will spout from the nozzle. It is only through similar joints that the operation of the kidney is revealed. If one breaks down, the colleague will never be convinced.

Will the colleagues be told everything or nothing? Compromise! Everything that is routine can be left out. So can secret recipes. There is no point in explaining that one must wear cotton gloves when weighing the kidney fragments so that sweat from fingers does not upset the measurements. These are little habits that we learn in good laboratories, just like those little secrets that are taught by great chefs. However, each time the absence of a precaution turns the hamster kidney into soup and causes a cloud of obscure numbers to come pouring out of the end of the hose, the Materials and Methods section should indicate that the relevant precaution has been taken. Although the style used in this section seems forbidding, behind every good adjective and noun hides a sharp-eyed critic for whom these are passwords. Reassured, he allows the author to continue to the next precaution just as the guards manning a series of checkpoints at the entrance to a fortress allow a caravan to enter. If the author makes it all the way through, the numbers inscribed in Table 4.1 will be accepted and the reader will finally start discussing their meaning.

According to Nietzsche, the desire not to want to deceive others is the same as the desire not to want to deceive oneself. It is the author that imposes upon himself the most gruelling of trials. For example, the article contains a little-used section titled 'critique of the methodology'. This section is a self-critique.

[7] It seemed appropriate to standardise the results with respect to the cortex sample because:

(a) some plasma samples were taken under poor conditions especially for the shorter time intervals where the delay between kidney removal and blood sampling could be an important source of error;

(b) cortex samples were treated in exactly the same way as those from other regions. They are therefore subject to the same causes of experimental fluctuation as the other samples.

It would have been nicer to relate the specific radioactivities of the isotopes in the kidney to their values in the plasma (which is what almost 'everyone' does!). But the plasma values which correspond to the shortest time intervals are missing from the table. No doubt they exist in the lab notebooks but they 'make no sense'. To deal with these results one must omit them as if they were the product of an unfortunate fluke. *Felix culpa* . . . this tell-tale accident allows one to 'standardise' the results in a prettier way. It allows one to smooth out the curves more successfully since the presumed sources of error compensate for one another as is indicated by point (b). This sort of self-critique most certainly contributes to the authors' honourable position. The reader will trust the results because the author anticipated not only his objections but also answered in advance questions of which the reader would not have dreamed. Who could now doubt the author or his overprotected results?

If anyone was still suspicious, he would not yet be referred from the texts to the actual laboratory practices, but rather from the text to others. These are only shown to close friends or reopened during investigations into rare cases of fraud. The lab notebooks and experimental records hold details of all the precautions taken, all the failures, and all the red herrings pursued. Someone in doubt over the numbers and the experiments could dive further into the book-keeping to prove that a certain experiment actually did take place, that a certain lab assistant was present that day, or that the animal keeper had in fact taken the water bottle away from the hamsters in order to put them in oliguria. The doubt will subside long before the stack of written traces runs out.

The more colleagues there are, the more the traces proliferate and pile up. If the cheat wants to conceal his cheating, he will even have to falsify his protocol books more and more cleverly (Bastide, 1981a). He will be prevented from fraud because of the extreme suspiciousness of his fellow colleagues. Actually he will have to be so clever that he will become, yes, an honest researcher just like the others.

EFFECT 6: TRIAL BY ORDEAL

[8] Pierre and Marie Curie (PMC): 'Here is the new hero emerging from this mixture, pitchblende, you see? It makes the air become conductive. You can even measure its activity with the instrument that Pierre devised, a quartz electrometer, right here. This is how we follow our hero's fate through all his ordeals and tribulations.'
Objector (O): 'This is not new, uranium and thorium are also active.'
PMC: 'Yes, but when you attack the mixture with acids, you get a liquor. Then, when you treat this liquor with sulphuretted hydrogen, uranium and thorium stay with the liquor, while our young hero is precipitated as a sulphuride.'
O: 'What does this prove? Lead, bismuth, copper, arsenic, antimony, all pass this test as well, and are also precipitated!'
PMC: 'Ah!, ah! But, if you try to dissolve them all in ammonium sulphuride the active something resists . . .'

O: 'OK, OK. It isn't arsenic, nor antimony, but it might be any one of the old heroes, lead, copper or bismuth.'

PMC: 'Impossible, my dear, since lead is precipitated by sulphuric acid while the hero stays in solution; as for copper, ammoniac precipitates that.'

O: 'So what? This means that your so-called "active substance" is simply bismuth. It adds a property to good old bismuth, that of activity. It does not define a new substance.'

PMC: 'It does not? Well, tell us what will make you accept that there is a new substance?'

O: 'Simply show me one trial to which bismuth reacts differently from your "hero" . . .'

PMC: 'Simply heat sulphur in a Boheme tube, under vacuum, at 700°C. And what happens? Bismuth stays in the hottest area of the tube, while a strange black (soot) gathers in the cooler areas. This is more active than the material with which you started. And you know what? If you do this several times, the "something" ends up being 400 times more active than uranium!'

O: . . .

PMC: 'We therefore believe that the substance we have extracted from pitchblende is an hitherto unknown metal. If the existence of this new metal is confirmed we propose to name it polonium after Marie's native country.'

When sent from the text to the laboratories, from the paper world to the material world, the scientific objectors are used not only to give solidity to the argument. They are also used to define new objects. What are these famous 'things' which are said to lie behind the texts? First, a list of victories. 'Polonium' won a series of trials in the Curies' laboratory at the École de Chimie before it becomes a 'thing out there'; it defeated uranium and thorium in the sulphuretted hydrogen game; it defeated antimony and arsenic at the ammonium sulphur game; and then forced lead and copper to throw in the towel; only bismuth went all the way to the semi-final, but it too was beaten during the final game about heat and cold! At the beginning of its definition, the 'thing' is a score-board for a series of trials. Some of these trials are imposed on it either by scientific objectors and tradition – to define for instance what is a metal – or tailored by the authors of the paper – like the last game about heat and cold. The point is that, at its inception, a thing is nothing but this list of scores.

The 'things behind' the scientific texts are thus similar to the heroes of a fairy tale. All of them are defined only by their performances: defeating the ugly seven-headed dragon; resisting precipitation; saving the King's daughter against all odds; finding one's way out of the tortuous kidney ducts; following Ariadne's thread out of the labyrinth, and so on ... Each of these performances defines what the hero does; in the beginning there is no other way to know his essence. However, this does not last long because each performance presupposes a competence which, from the start, retrospectively explains why the hero withstood all the ordeals. The hero is no longer a scoreboard of actions; he is an essence slowly unveiled by each of his manifestations. Once polonium is accepted as an element the meaning of the list above changes dramatically: before, it was what made the new polonium exist inside the laboratory of Pierre and Marie Curie; afterwards, it becomes the agent that revealed the essence of an everpresent substance observed by the Curies.

EFFECT 7: THE FATE OF A CLAIM IS IN THE HANDS OF THE READER

[9] Evidence for counter-current water exchange in the inner regions of the hamster kidney.

[10] 'If François *et al.* have shown a counter-current water exchange, they did not provide a physiological explanation for the dissipation of the osmotic pressure gradient at the onset of diluted urine production. The present article shows measurements of the rate of water renewal under conditions of osmotic diuresis induced with mannitol.'

[11] 'François *et el.* claim to have shown a process of counter-current water exchange but their experiments are based on the use of hamster kidneys whose papilla are very long. It is precisely this characteristic that prevents us from extending their results to the kidneys of other mammals.'

[12] 'Shown? Shown? Well that was quick. OK, so they cut up kidneys. They sprinkled on some radioactivity which probably made their hands peel. But what does that prove? Have you read their articles? There's a difference between two curves and that's all there is. If you ask me, the rest is nothing but rhetoric. So far as hamsters are concerned, they don't get much of a look in at all.'

[13] 'Mammal kidneys possess counter-current water exchange that prevents the dissipation of the sodium gradient which serves to concentrate urine.'

The scientific article is read by other writers who cite it, discuss it, ignore it, suppress it, or believe it. Paragraph [9] describes what the authors claim to have accomplished. But will it be believed? The history of science is no more predictable than history itself. The outcome depends on gathering other authors/readers, and on what they find to disagree with.

The author of paragraph [10] believes what our physiologists have said so that he can move directly on to another problem: how the gradient of osmotic pressure vanishes. The more credible an article is, the more rapidly it can be redeployed to deal with another question. As in paragraph [2] where the authors claim 'decisive advances' for themselves, the author of paragraph [10] suggests that the others have not found all the elements of the solution.

Paragraph [11] is even more cruel. The article is not challenged but the extension that the authors could have claimed if no other colleagues had been present, is denied. The problem of induction is not as important as it appears. Without [11], nothing would have prevented François *et al.* from passing effortlessly from hamsters to all mammals. Conversely, if the competitors had been fiercer and our physiologists less well armed, they could have reduced the article to forty slices from five hamster kidneys! There is no limit to the degree of induction. It depends on might not right.

The person interviewed in paragraph [12] is a boor who denies any relationship between François's article and the hamster kidney. From this excerpt, it is apparent that relativism and the problem of reality is no more a philosophical question than that of induction. It is rather a trial of strength to be settled in action! If [12] is true then [9] is but an artefact. If [9] is true then [12] is nothing but the exaggerated bluster of a half-wit. On the other hand, paragraph [13], which is taken from a textbook, concedes so much reality to sentence [9] that there is no indication that it may be disputable or that someone has actually discovered it. That which is real and that which is not depends on each author, or rather, on what each author does with all the articles that preceded his own. The final degree of conviction is thus a mass effect.

With each text the reality of all those who are cited or ignored either grows or diminishes. There is no point is discussing an 'exact representation' of the operation of the hamster kidney. This appears at the end of a battle when everything has been played out. The 'exact representation' is the particular literary effect produced by the textbook. In scientific literature, 'demonstration' is a rare alignment of procedures, gestures, inscriptions and paragraphs that advances as slowly as the monstrance at the Corpus Christi procession. By going from five hamster kidneys to the 'operation of kidneys in mammals' and then back to the particular hamsters sacrificed above, a representation is displayed.

The representation of reality is one result among others, built sentence by sentence in laboratories and scientific articles which can be deconstructed sentence after sentence by the sociologist or semiotician who reads scientific articles: *quod erat monstrandum*.

NOTES

1. A preliminary and uncorrected version of this article has appeared without the authors' final approval in *Études Françaises*, 19 February 1984, pp. 114–33.
2. A few historians of science have been interested in the ways scientists were writing, and especially rewriting, their texts: for instance Koyré (1968) on Newton, Grmek (1973) on Claude Bernard, Feyerabend (1975) on Galileo. What a few scientists had said, that science was a form of writing fiction and not a direct rendering of Nature (Medawar, 1964) has been confirmed by discourse analysts (Gilbert and Mulkay, 1984; Woolgar, 1976, 1980) and by semiotic studies (Latour and Fabbri, 1977; Latour and Woolgar, 1979; Bastide, 1979, 1980, 1981a, 1981b, 1981c). Laboratory studies have, of course, encountered the problem of papers and paper construction (Latour and Woolgar, 1979; Knorr-Cetina, 1981; Lynch, 1985; Law and Williams, 1982) integrating this problem in the precise setting of laboratory practices. Two other notions are germane to these studies: that of rhetoric (Perelman, 1970; applied to science by Gusfield, 1976), and that of style (Granger, 1968). A few linguists have studied the structure of scientific texts (Gopnik, 1972; Lofflet-Laurian, 1980). Several specialists coming from literature studies have found the scientific text interesting as a genre (Bazerman, 1981, 1984; Yearley, 1981). The study of the visual aspects of scientific literature is less developed but there are some interesting exceptions (Ivins, 1938, 1973; Goody, 1980; Ong, 1982; Jacobi, 1982). Most of them are gathered or reviewed in Latour, 1985. Finally, there has been a mounting interest in scientific literature at the junction of the history of the publishing industry and that of science (Eisenstein, 1979; Meadows, 1980; Anderson *et al.*, 1983).

5 The Heterogeneity of Texts

JOHN LAW

INTRODUCTION

The text is the secret weapon of science. It is sent out from the laboratory and, if it does not strike terror into the hearts of those who read it, at least they are often obliged to take it seriously. By virtue of its transportability, its durability and its structure, it is often able to operate as a relatively autonomous agent thousands of miles from those who sent it out.

In Chapter 4, Latour and Bastide examined some of the ways in which the text obtains its power. Their approach was to create what may be called a 'semiotic reader', a fictional reader who is interested in the text but who is also radically sceptical. The effects that they described may be seen as textual methods of translation that are designed to force this reader into the laboratory in the pursuit of scepticism. They are, in other words, designed to prevent a reader who is interested in the argument from picking holes therein. However, as all those who set pen to paper know, this is only half of the battle. Before the sceptic can be convinced, it is first necessary that he or she should be *interested*. The reader has, in other words, to be sucked into the series of translations that will, if scepticism is pressed to its ultimate extreme, lead back into the laboratory. The problem, then, of making the paper attractive, of finding a journal that will publish it, of discovering or building a readership, is crucial both to individual scientific success and the capacity of certain texts to exercise power at a distance. It is this question that is addressed in the present chapter.

This problem is approached by considering an empirical example and looking at the way in which a particular group of scientists have sought to build an actor-world via the agency of a particular text. The argument falls into two main parts. First there is the question of *where* scientists choose to send their texts or, more precisely at *whom* or *what* the texts are aimed. This requires that we step outside the text itself and consider negotiations about the journals to which texts might be submitted, negotiations that are organised around conceptions of the interests of possible audiences. Second there is the question of the construction of the text, of the forces to be borrowed and the way in which they should be arrayed on the printed page. Here, though the analysis is more closely related to that of Latour and Bastide, the focus is once again on the way in which particular scientific elements that might, in the abstract, be thought of as uninteresting, are in fact charged with interest by the juxtaposition of appropriately chosen terms. It is argued that the forces these translate are entirely heterogeneous in nature. The 'scientific', the 'technical', the 'social', the 'economic' – no matter how pure the article may appear to be at first glance, all are intermingled in an attempt to build an argument that will enrol and operate on the presumed interests of the reader.

The argument, then, is that words, which appear so insubstantial, in fact index forces of different types and operate to restructure the environment of science and technology in a variety of ways. In particular they operate to work upon the interests of the reader and force a 'route march' past a particular point that is held to be of key significance. In the case of an empirical paper this takes the form of specific findings. The aim, then, is to endow this key point with the capacity to act far beyond its place of origin – to collapse the distinction between inside and outside on the one hand and 'macro' and 'micro' on the other.[1]

TRANSLATING AUDIENCES

The scientific paper presents itself as an actor-world, an actor-world which operates upon the reader to translate him/her to a particular place in that world. It seeks, in other words, to enrol the reader by means of an appropriate array of borrowed forces. On the other hand the paper never operates in a vacuum. Texts arriving from distant parts are received as a function of the structure of forces obtaining locally. To put this in another way, a text is only strong if (a) its components have force at the point where the text is received and (b) they are recognised as having been properly borrowed and juxtaposed. In other words, the scientist or technologist who wishes to influence a reader has two linked problems. First, a reader must be identified. And second, the appropriate array of forces has to be found that will translate or enrol that reader. The two are interrelated because what count as 'appropriate' words are a function of the characteristics of the reader. A set of forces that might operate upon one reader will naturally leave another quite unmoved. In this section the first question is considered – that of the selection of the reader.[2]

Selection of a reader mirrors the selection of a subject matter in an obvious sense. Very often the selection of a reader comes first and then the appropriate subject matter is determined subsequently. In the case discussed here, though there were other reasons for the particular choice of subject matter (the effect of a particular family of synthetic polymers (DIVEMAs) on the rate of absorption of substrate by cells (pinocytosis)), there is no doubt that the biochemists were aware of the relationship between subject matter and audience. Thus the director of the laboratory spoke in the following terms to Williams early in the project:

> We could have chosen all sorts of polymers to do this experiment with, but having chosen one that other people have done things with will in fact make it much easier to get it published. You can start by saying 'so and so has said this, etc., etc. and it's interesting to find out whether DIVEMA does so and so'.

Indeed, the notion that this or that topic should be of interest to this or that person, group or class of persons, seemed to organise much of the discussion about how and to whom the paper should be presented. If a topic was interesting to another then he or she might be attracted to it. Correspondingly, other actors were endowed with interests which led them to be differentially attracted to different topics. Talk of interests and what might be interesting seemed, then, to index a set of forces – attractions, lack of attractions, repulsions – that were presumed to exist in the field. It seemed, in other words, to be a vocabulary for talking of the degree of feasibility of a given translation. If one object was attracted to another then, so long as the latter was available for deployment, the former became a candidate for translation. Thus consider the following conversational exchange between the two most senior members of the team:

> 'We wrote to you (that the article could appear in) *Cancer Quarterly*. I just don't know. We're a bit uncertain about this. We just don't know who would be interested in this. Do you think that it ought to go in some sort of polymer kind of journal?'

> 'It would be easy to get it into the journal here in Stiftungburg, or *Fundamenta Polymer*. [B] published some papers in this. It's not uninteresting. On the other hand the special effect – especially if you would like to keep your title – the people in the other area would be interested. So *Cancer Quarterly* sounds good to me.'

The two journals in question are seen, at least by the second speaker, as being susceptible to the attractions of the article. To put this another way, they are seen as being susceptible to its force – or more precisely, the forces that it itself has borrowed by means of juxtaposition. The borrowed forces of the article are seen as being capable of enrolling the journals in question – of moving them from an initial state of ignorance to a willingness to publish the article – *because they are in turn juxtaposed with an existing array of forces native to the journals and their readers*. The durability of the article thus depends not only on the force that it attempts to array in the text but also on whether or not it can successfully borrow and enrol them. Hence the concern with which scientists view the question of audiences. Obviously enough, the wrong audience can take apart an array of forces and reject the argument of an article. It can refuse the place proposed for it in the actor-world of the article. Hence the concern with who is interested in what, a concern for the environment within which the borrowed forces of the emissary will have to work. The overall aim is to juxtapose and enrol. Some forces may be transmitted into the distance, but for this to be successful they have to operate upon and not against their indigenous cousins. It is this fact that defines the structure of the text – a structure that is accordingly quite different when the intended audience is formed of grant-giving agencies or patent attorneys.[3]

NOTES ON THE CONCEPT OF 'INTERESSEMENT'

At this point is is necessary to introduce a novel term – that of 'interessement'. This neologism taken from the French, may be defined as the action of interesting, enrolling or translating. It will be recalled that to translate is to speak for others, to make oneself indispensable, and thereby to determine and allocate roles and to decide what may be exchanged between the occupants of those roles. It is, in short, to impose a structure upon others. It will also be recalled that the theory of the actor-network assumes that there is no overall structure – that there is always a multiplicity of actor-networks each trying to impose its own structure on potentially unreliable entities and thereby borrow their forces and treat them as its own. We have talked much of the way in which successful enrolment requires the construction, by means of successful juxtaposition, of a locally overwhelming structure of forces. We have not so far, however, considered the elementary form of such enrolment. The concept of 'interessement' makes this possible.

Consider briefly again what is involved in 'interesting' readers or journals in an article. Imputed interests must be mobilised. Audiences must be persuaded of the interest of the piece. They must be attracted to the piece. They must be enrolled. Given the fact that there are many articles submitted to journals, and the fact that the volume of published papers is quite beyond the capacity of any one individual to read, there is inevitably an element of competition. Journals and readers have to make choices between papers. They have to decide which are the most attractive. The question, then, is which paper will successfully mobilise the forces necessary to enrol the audience in question? Which actor-network will succeed in structuring this part of the social world? In any given case the outcome may be a more or less foregone conclusion. However, as the study of the electric vehicle suggests, the scientific and technological world is full of struggling actor-networks. A run of successes does not preclude the possibility that next time another actor-network will array such a powerful structure of forces that one finds oneself localised, unable to publish, perhaps with one's forces so scattered that it will be difficult to publish anything using, for instance, one's favoured method ever again.

Any given attempt to enrol is likely to be complex. There will be many different actor-networks, some in competition, others in at least tacit cooperation, using the links that others have established in their attempts to translate. However, the elementary situation is quite easily described. Just as we may picture two papers competing with one another for the attention of an audience, we may picture two actor-networks competing with each other for the affiliation of a third. *The elementary case of translation is thus triangular*. One entity enrols. Another is enrolled. And a third fails in its attempts to enrol. This we call interessement. As can be seen from Figure 5.1, *interessement involves attracting one entity by coming between that entity and a third.*[4] The case of interests and interesting articles discussed in the previous section thus counts as a case of interessement. The latter is, however, much more general in scope. Interessement may indeed operate through the medium of interests, but also via naked force, apparent inevitability, love, duty, bribery, physical fact, superior command of information, spirituality or lack of imagination. Thus, though scientific and technical papers appeal to their reader's interests, they do far more than that.

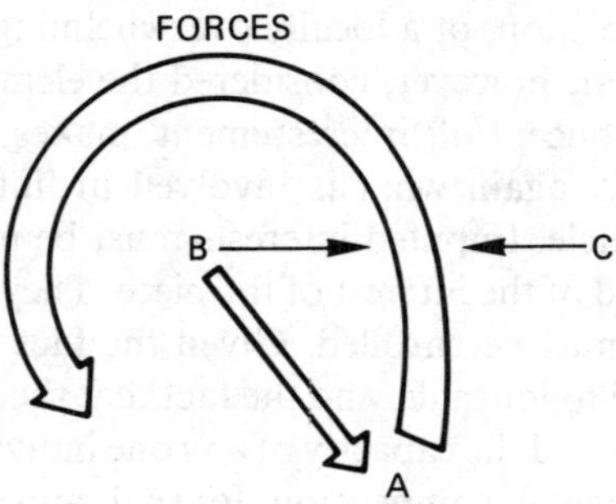

FIGURE 5.1 The elementary form of interessement

FROM GENERAL TO PARTICULAR: THE STRUCTURE OF TEXTUAL INTERESSEMENT

Just like the laboratory the text is an attempt to assemble forces in a powerful combination. Just like the laboratory, these forces can escape and go native. If they do this they leave the words high and dry. The text is no longer worth anything more than the paper it is printed on. The problem is to harness those forces by deploying words in an appropriate array and then to stop them escaping. The solution is to attempt a series of interessements. The translations of scientific and technological articles may thus be treated as a series of interessements.

Consider first the fact that for the reader it usually arrives in the neatly packaged form of a journal or an offprint from a journal. Already, then there is a context of potential forces – the forces borrowed by the title of the journal. 'Everyone knows' that lots of poor papers are published in journals, but nevertheless its title carries some kind of weight, particularly if it is *Nature* or *Science*. Then there is a title of the article, a list of authors, an indication of their institutions of affiliation and, perhaps at the end, a section acknowledging the help and support of funding institutions or fellow scientists. The title represents a set of heterogeneous forces, though these look 'scientific'. Thus, though the biochemists studied in this chapter called their paper 'The failure of DIVEMA to stimulate pinocytosis',[5] this conceals a wide range of social and technical elements whose nature becomes somewhat clearer as one passes through the text. The title conceals, but more importantly from the standpoint of translation, it provides a context for what is to follow. So too does the list of authors, of institutions and of bodies acknowledged, though here their heterogeneity is manifest. Words stand for people, universities, laboratories and voluntary and government agencies – all of these feature in the text. This does not mean, of course, that the force of all these entities will necessarily be successfully borrowed. Cases of fraudulent use of titles are occasionally uncovered. However, the assumption is that these forces are indeed properly appropriated unless it is proven otherwise. And overall the effect of this 'packaging' is twofold. First it offers the authors the provisional right to be heard because it makes it clear that they are serious researchers supported by serious bodies and employed by serious institutions. This contributes to interessement because it tends to disconnect the reader from the alternative of dismissing the text without even reading it on the grounds that the authors are simply charlatans. Such headlines thus punctualise a potential readership.[6] But second, and just as important, it makes a link between the range of important institutions mentioned above and such apparently scientific terms as DIVEMA and pinocytosis. This has the effect of charging the latter with a provisional force which they might otherwise have lacked.

The paper written by the biochemists is, like many papers, divided into a number of standard sections. There is a brief 'Summary', an 'Introduction', a section entitled 'Materials and Methods' and a couple of pages of 'Discussion'. There is also a long list of references – twenty-nine in all. The purpose here is to illustrate the claim that a paper may be seen as a sequence of interessements in which heterogeneous forces are marshalled to move a wide range of readers through a particular series of positions, so there will not be an exhaustive analysis of the article in question. Attention is concentrated primarily (though, even then, not exhaustively) on the 'Introduction'. This is reproduced in full in the appendix at the end of this chapter.

The first point to note is that this provides a broad context within which the work to be reported in later sections can be placed. Certain terms are deployed with no explanation or justification. Others are defined in terms of these, or each other. Thus the term 'synthetic anionic polyelectrolyte' is undefined but is used to locate 'divinyl ether-maleic anhydride', 'DIVEMA' and 'pyran copolymer' (which are said to be synonyms.) Again, 'interferon production', 'macrophage activation' and 'tumour regression' are undefined, help to specify 'physiological responses' and, in particular, are attributed to the activity of DIVEMA.

Thus a network of terms is built up that purports to represent a variety of forces. In particular, DIVEMA is endowed with a set of forces which relate to its chemotherapeutic qualities. The structure of the interessement is clear. The reader needs to know why he/she should be attracted to DIVEMA. Why, in other words, he/she should not turn instead to some other paper. The answer is that a polymer with such potent anti-cancer effects is worthy of consideration. Chemotherapy, then, comes between the reader and his/her indifference or hostility to DIVEMA. It lends its force to DIVEMA and simultaneously binds the reader to the text. The assumption, of course, is that the reader is bound to be interested in the treatment of cancer. If this assumption is faulty then, short of an interest in DIVEMA itself, there is little about the first paragraph to attract the reader. Thus, as noted earlier, the text needs to find forces at its point of reception which it may work upon and translate. If it is unable to do this then it will fail – or will be seen as empty words.[7]

The first paragraph, then, constrains the reader in so far as DIVEMA is rendered attractive by its borrowed chemotherapeutic forces. The second paragraph attempts another translation: it tries to move the reader from an interest in the anti-cancer properties of DIVEMA to an interest in its pinocytic properties. This interessement succeeds to the extent that it is able to lend the forces of chemotherapy to the action of pinocytosis and thus prevent the reader from saying: 'well, I'm interested in the chemical treatment of cancer cells by DIVEMA but I can't see what that has to do with pinocytosis.' In sum, interessement has to link chemotherapy to pinocytosis and disconnect the reader from lack of interest in the latter. The way in which this is attempted is first to assert that, given the interest in the anti-cancer effects of DIVEMA, it is important to clarify the mode of operation of the latter at the cellular level. This claim, which is certainly not self-evident in nature, trades upon the assumption that clarification of the DIVEMA–chemotherapeutic link is (a) important and (b) synonymous with a study of DIVEMA–cell interaction so that if chemotherapy is to be understood, the behaviour of cells must be investigated. Thus we witness a translation from the clinical to the cellular. Second, two ways in which molecules may influence cellular behaviour are specified – by binding to the cell membrane, or by internalisation. The assumption is that there are no further options and that an interest in cellular activity is appropriately specified in one or other of these two ways. Third, it is stated that the only way in which 'macromolecules such as DIVEMA' can enter cells is by pinocytosis – phagocytosis, for instance, being impossible. In this way the force of chemotherapy is lent to the study of cell behaviour, the borrowed force of the latter is lent to the study of internalisation and the combined forces represented by this are in turn lent to pinocytosis.[8] This is the array of forces that here constitutes the interessement – that intervenes between the reader and his/her potential lack of interest in pinocytosis.

However, this is not the last attempt at interessement to be found in the introduction. Another translation is necessary – the move from an interest in the effects of DIVEMA upon pinocytosis to an interest in the results obtained when DIVEMA is tested 'using two *in vitro* model systems'. Once again, then, an equivalence is proposed between two objects that are by no means self-evidently identical, and once again forces are marshalled in order to defeat the possibility that the reader will escape from the proposed translation. This time, of course, the danger is that he/she, while conceding the importance or interest of studying the effects of DIVEMA on pinocytosis, might find that the rat visceral yolk sac and rat peritoneal macrophage systems only inadequately model this. Interessement takes the form of a bold claim – that 'both systems were considered very suitable' because they had previously been used to study phenomena ('the kinetics of pinocytic uptake of macromolecular substrates' and 'the effects of potential modifiers of pinocytosis') which may be seen as similar if not actually identical to the phenomenon to be considered here. If the reader shares the implicit view of similarity then this hurdle too has been cleared. The reader has been moved from an interest in chemotherapy to an interest in the behaviour of rat yolk sacs. Though this movement, if considered in the abstract, is really quite remarkable, for precisely this reason it makes an excellent example of the operation of translation. A point of obligatory passage or a problematisation has been built.

Even so, the series of introductory translations is not yet complete. The operation of the model system remains to be explained. Thus it is not good enough to show in general that (for instance) the rat yolk system can model pinocytosis. It is also necessary to show that the *particular* experiments undertaken are proper examples of that model system at work. In other words, it is necessary to counter the danger that the reader may, even then, escape by arguing that, though the model system is potentially a proper way of studying the effects of DIVEMA on pinocytosis, the actual experiments carried out were in some way inadequately performed. It is not intended here to spell out the precise form of this particular translation which moves us through the last paragraph of the introduction and into the section entitled 'Materials and Methods'. However, it is worth noting, in brief, that the choice of DIVEMAs, their origins and their mode of preparation is explained with the apparent aim of defeating the sceptic who might say that they were badly chosen or synthesised and hence irrelevant to the question of the influence of DIVEMAs in general on pinocytosis.[9]

We have so far looked at the introductory section of the paper in question which leads, as has been indicated, from the general to the particular. Of course this does not, by itself, add up to a whole article. There is a section on results, but there is, in addition, one final ingredient necessary for the success of the paper: the construction of a series of translations that reverses the journey from the general to the particular. For just as the reader has accepted the salience of a table of results to a general problem, he/she must now be persuaded that these particular figures redefine the relationship between DIVEMA on the one hand and pinocytosis and chemotherapy on the other. And this is exactly what the discussion section of the paper does. It 'draws out' the 'implications'[10] of a set of figures by showing how they define DIVEMA–pinocytosis interactions. In part this involves establishing links between the two that had not previously been specified:

> . . . the rapid uptake of DIVEMA by macrophages must be attributable to a receptor-mediated process, a conclusion further supported by the observed competition between unlabelled methoxy-aniline-substituted DIVEMA and ^{125}I-labelled DIVEMA.

In other cases reconstruction is involved, since the 'implications' of the experimental results contradict translations that have previously been made:

> . . . the only correlation between our data and those of [B] *et al.* [23] was the inhibition observed with high molecular weight species.
>
> Contrary to expectations from some earlier data, there was no enhancement of pinocytosis and in some case inhibition was seen.

With this the argument of the paper is complete. The circle has, so to speak, been squared. The next section considers a number of important characteristics displayed by this series of translations and the words that are used to build them.

THE FUNNEL OF INTERESTS

It is worth underlining a feature of the introduction that has already been indicated: namely that as one passes through the sequence of translations there is a move from the general to the specific. 'Rat visceral yolk sacs' may signal an array of forces that have potency in certain places, but it is a safe bet that 'tumour regression' and 'therapeutic potential' represent forces with a much greater potential capacity to mobilise or enrol. Since the paper is an array that is designed to maintain its durability at a number of distant sites, it clearly makes excellent sense for the authors to try to deploy forces that are likely to be found in a wide variety of locations. Despite the optimistic note sounded in the first quotation on p. 69, DIVEMA does not have the same pulling power as chemotherapy. Yet it can be linked to chemotherapy and the force of the latter may be borrowed by DIVEMA and the experiments thereon. This principle – that of starting with a force that is generally respected and thence of translating the reader from the general to the particular – is one that is elsewhere called the 'funnel of interests.'[11] As a principle in the organisation of persuasion it seems to be general. The mouth of such a funnel is broad, in order to suck in as wide an audience as possible. Thereafter it narrows as the broad range of interests is concentrated and focused on a particular series of points. Interests are thus channelled and flow along a course that is determined by the linked series of interessements. These are deployed by the authors in order to prevent escapes – to ensure that the reader reaches the approved conclusion and then moves with this back to the general once more.[12]

There is, however, another way of presenting the operation of the series of translations that exist in this text. This is to note that they establish a direct link between problems to do with DIVEMA in chemotherapy on the one hand and the problem of the particular DIVEMAs on particular cells as measured by particular instruments in a particular system in a particular laboratory on the other. The consequence is that an actor-world has been built in which apparently insignificant events in one laboratory have an extraordinary capacity to resolve general problems. A situation has been generated in which local phenomena may borrow forces far beyond their wildest dreams and exercise influence in far distant places. It is suggested that this is a principle on which most if not all texts work. They are designed to show that a great deal hinges around very little. They are reductionist. They oblige the reader, who is interested in the solution of this or that *general* problem, to move through one *particular* point which has been endowed with the power of solving or specifying that general problem. Here the results of a set of very special laboratory experiments resolve the problem of how DIVEMA affects pinocytosis and operates in chemotherapy. Readers find themselves accepting that it is not the behaviour of a particular batch of rat visceral yolk sacs on a particular day with a particular derivative of DIVEMA that is at issue, but rather a whole established wisdom about DIVEMAs and their role in relation to both pinocytosis and chemical therapy.[13] These experiments are constituted, then, as the key to problems that arise in other laboratories, indeed in hospitals between doctors and patients up and down the land. The pattern discovered by Callon for the case of the electric vehicle is reproduced precisely here. A series of hierarchical *problematisations* has been erected in which solutions to the more general depend upon a solution to the most particular. Everyone has to pass through one laboratory. In sum, a short circuit has been created between the particular and the general (see Figure 5.2), a short circuit that depends upon a set of translated equivalences.

The solidity of this chain of equivalences does not, however, rest exclusively on the biochemical facts and figures arrayed on the page for, as indicated earlier, references also pepper the text and citation constitutes an integral part of the process by which forces are borrowed and translated to push the reader along. Other people, laboratories and their papers are borrowed to add weight to this process of channelling. Though the main text deals predominantly with physical or biological objects (for instance, rat yolk sacs and ^{125}I-labelled PVP) and physical processes (e.g. 'kinetics of pinocytic uptake') and only occasionally introduces people, laboratories or other organisations, these form an integral part of the array of forces and are routinely borrowed, in particular at those points in the argument where the reader might be expected to treat a statement of fact as lacking force and thus escape from the channel laid out. For instance, the authors apparently feel comfortable with a bald statement of the kind 'Divinyl ether-maleic anhydride . . . *is* a synthetic anionic polyelectrolyte' (emphasis added), but feel it necessary to introduce other authors and papers when they write 'Divema has been reported to cause "activation" of macrophages [11, 12]'.[14]

However, references are only the tip of the iceberg for even if the main text alone is considered and headings and acknowledgements are excluded, scientific and technical facts, phenomena, methods and processes all, as suggested above, rub shoulders with people, laboratories and companies. These, in turn, interact with such entities as

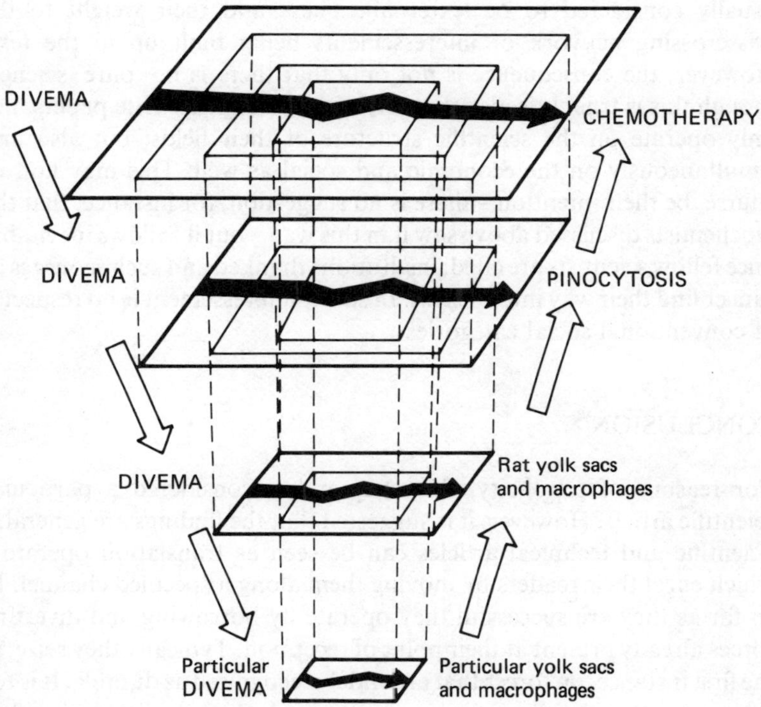

FIGURE 5.2 Linked translations or the funnel of interests

tumours, toxic side-effects and clinical trials. It is being argued, then, that there is nothing 'pure' about the traces that constitute this article – just as there is nothing pure about any paper that cites other authors and so situates itself in terms of people. *The text both constitutes and indexes a mixture of 'scientific', 'social', 'economic' and 'organisational' forces.* In order to add to the solidity of the funnel of interests, and to push the reader along to the conclusion that particular experiments have relevance far beyond the time and place that they were undertaken, heterogeneous elements are regularly assembled by the authors of scientific texts. This is a general finding. Texts in science are *constitutively heterogeneous* in nature and scientists operate not only upon the 'internal' scientific world but simultaneously upon the 'external' as well. This is because the strength of an argument, whether scientific or not, rests upon the number and power of the elements that may be borrowed to give it force. In science force is obtained, in part, from factors that are usually considered to be 'external'. These add their weight to the crisscrossing network of interessements being built up in the text. However, the consequence is not only that there is no 'pure' science, though this is true. It is also that scientists, when they write papers, not only operate on the scientific structure of their fields, but also and simultaneously on the economic and social as well. This may not, of course, be their intention – there is no suggestion, for instance, that the biochemists discussed above saw it in this way – but it follows inevitably once fellow scientists are cited, institutions thanked and such scourges as cancer find their way into the text. In short, interessement is no respecter of conventional social categories.

CONCLUSION

For reasons of simplicity, this chapter has considered a particular scientific article. However, it is suggested that the findings are general.[15] Scientific and technical articles can be seen as translation operators which enrol their readers by moving them along a specified channel. In so far as they are successful they operate by borrowing and diverting forces already present at their point of reception. Typically they seize, in the first instance, on forces that are widely recognised as durable. It is for this reason that their structure is appropriately described as a 'funnel of interests'. The introduction of the scientific paper is broad as it is designed to operate upon as wide an audience as possible. The funnel narrows as the widespread and durable forces that constitute its mouth are borrowed by more local forces and the paper starts to transmit the reader from the general to the particular by asserting their equivalence.

The chain of translations that makes up the funnel is constituted by a series of linked interessements. Interessement is defined as a triangular relationship in which one entity enrols a second by coming between this and a third. The problem, then, is to find the force that is needed to stop the second affiliating to the third and secure it firmly to the first. These forces take various forms. Interests are important, but only constitute one possibility. Violence, depictions of fact, displays of wealth, manipulation of theory and appeals to duty – these are some of the other forces deployed in the attempt to ensure interessement. Obviously successful translation depends on locating forces already active in the environment, and then attaching them by means of interessement. Thus a good scientific paper, like a good general staff, makes use of the configuration of forces already available, and bends these to its own ends. The scientific paper does not use violence, but despite its apparently 'pure' format the forces it actually deploys are heterogeneous in nature. In the paper analysed in this chapter, people, grants, diseases, laboratories and official agencies all made their appearance in addition to findings and theories.

The words of a scientific paper may thus be seen as a *network of problematisations* which stand for an actor-network, define the problems that must be overcome, the way these should be tackled and the location where this is to be properly attempted. Therapy, tumour, DIVEMA, toxicity, molecular weight, pinocytosis, macrophages, hydrophobicity, positive charges: all these words and others besides represent the structure of the actor-world built in the paper. Indeed, that world may, for certain purposes, be treated as a string of forceful words which represents the laboratory as an obligatory point of passage. In the previous two chapters some of the methods by which these forceful words are assembled and deployed have been described. Though such analyses are essential they are labour-intensive, time-consuming and expensive. However, quantitative and semi-automated analyses of such *networks of problematisation* lie within the realm of possibility. Studies in depth can thus be complemented by studies in breadth. The method of co-word analysis which renders this possible is described in the third part of this book.

NOTES

1. These distinctions, though they have wide currency in social life are accordingly avoided in actor-network analysis. For the argument applied to laboratories see Callon and Latour (1981), Latour (1983) and Chapter 3 of this volume. For the argument in the context of organisations see Law (1984a).
2. The data are drawn from a study by Rob Williams and myself of a group of biochemists. I am grateful to both him and them for permission to reproduce this material here. Other ethnographic details are to be found in Williams and Law (1980); Law and Williams (1981, 1982); Callon and Law (1982); Law (1983); and Williams (1984).
3. For a discussion of the very different types of texts produced in these two cases, see Chapter 6.
4. The etymology of the term, inter-esse, is very obvious for the triangular figure.
5. In order to maintain some degree of confidentiality for the biochemists studied it was agreed to conceal identities. Accordingly, all names of individuals and organisations have been changed, and it has been necessary

on occasion to alter other data. Where the latter has been necessary this will be indicated. Here the title is not fictional but is derived from an unpublished paper rather than the published text discussed in this chapter.

6. There is much valuable work in ethnomethodology on the importance of openings for the subsequent proper reading of a text. For an example of an analysis of a scientific text see Woolgar (1980).
7. Readers will note that there are a couple of sentences that describe a version of the recent history of chemotherapeutic practice with DIVEMA. This can be seen as another, subsidiary interessement, designed to cope with the escapee who might say, 'Well, I know that DIVEMA *has* been used to treat cancer, but it is now well known that its toxic side-effects make it quite unsuitable.'
8. It is conceded that binding to the cell membrane is also a legitimate possibility.
9. Also mentioned are the origins and preparation of radioactive markers, certain aspects of the experimental procedure, and certain features of the way in which results were calculated and presented.
10. The fact that I have placed these terms in quotes should not be taken to mean that I doubt the translations being proposed by the authors. They are well recognised as having an expertise in the study of pinocytosis. Rather the quotes are intended to indicate that there is nothing 'automatic' about such processes of argument. They are sustainable only in so far as an array of forces may be borrowed and tamed.
11. See Callon and Law (1982) and Law (1983).
12. The hydrodynamic metaphor is in many ways a helpful way of conceiving the action not only of texts, but also of speeches and sales pitches. These all have in common the fact that they maximise their appeal by chaining generally acknowledged forces to particular points of passage. Sometimes, as with pitches to sell life insurance, these funnels are constructed with, as it were, malice aforethought. I would not wish to imply, however, that the authors of the biochemical paper being discussed here had anything other than the best of motives. They were not simply suggesting that their work had relevance for chemotherapy. They firmly believed this to be the case and, since their belief was shared by (amongst others) the editors of the journal, they had good reasons for their views.
13. Latour (1984), in a magnificent study of Pasteur, makes essentially this point but is able to make it more strongly because Pasteur was able to construct a much broader funnel of interests than the biochemists in this study have so far managed.
14. Notice also the more cautious 'has been reported'. Latour and Woolgar (1979) discuss these 'modalities' at some length, but though there is obviously some relationship between the degree of confidence with which a finding is reported and its actual force, this is very far from perfect and cannot be relied upon as a short-cut way of determining the strength of forces.
15. I am not unaware of the fact that this constitutes a translation very like the one described.

APPENDIX

Divinyl ether-maleic anhydride (known either by the acronym DIVEMA or as pyran copolymer) is a synthetic anionic polyelectrolyte, which, following *in vivo* administration, initiates a wide variety of physiological responses including interferon production, macrophage activation and tumour regression [1,2]. In the 1960s DIVEMA was tested clinically as a potential anti-tumour agent, but trials were soon abandoned owing to the high incidence of toxic side-effects [3]. These early trials were carried out with DIVEMA samples that were polydisperse and it has now become apparent that toxicity of DIVEMA is confined to the higher molecular weight fractions [4]. Recently, interest in the therapeutic potential of DIVEMA has revived and a new clinical trial begun using well characterised material in the mol. wt. range 12000–15000 [5].

In the light of this renewed interest it is important to clarify the mechanism of action of DIVEMA at the cellular level. The interaction of DIVEMA with cells may be either intra- or extracellular. Binding of many molecules, including some hormones, to the plasma membrane is sufficient to trigger a cellular response, but in other instances a substance must be internalised before it can exert an effect. Macromolecules such as DIVEMA can only enter cells by pinocytic uptake and in the present study we have investigated DIVEMA as a substrate and possible modifier of pinocytosis using two *in vitro* model systems, the rat visceral yolk sac and the rat peritoneal macrophage. Both systems were considered very suitable for this purpose, as they have been used extensively to define the kinetics of pinocytic uptake of macromolecular substrates [6–8] and also the effect of potential modifiers of pinocytosis [9,10]. DIVEMA has been reported to cause 'activation' of macrophages [11,12], so that the investigation of the uptake of DIVEMA by macrophages is particularly relevant.

We have tested DIVEMA in three different molecular weight ranges as a potential modifier of the pinocytic uptake of two substrates, ^{125}I-labelled PVP and colloidal ^{198}Au. ^{125}I-labelled PVP is a fluid-phase pinocytic marker in both cell types, whereas colloidal ^{198}Au enters to a considerable extent by adsorptive pinocytosis, particularly in the macrophage [6,8]. We have also examined the effects of three derivatives of DIVEMA, in order to explore structure–activity relationships in this group of compounds: one contains a proportion of betaine-structure, and thus carries some positive charges; one is substituted with octadecane groups to confer some degree of hydrophobicity; and one contains pyrrolidine units, and so has a somewhat lower net negative charge. In addition, by using two different radio-labelled preparations of DIVEMA, we have measured directly its own rate of pinocytic uptake by rat peritoneal macrophages.

6 Mobilising Resources Through Texts

ARIE RIP

In the preceding chapters it has been argued that texts are powerful in science. This is because they have the capacity to act at a distance through the presentation of a carefully structured actor-world. They signpost the passageways that interested readers should follow, and they make these signposts forceful by providing traces that lead the reader both to the laboratory and to other scientific articles. Readers who wish to deconstruct the text – normally other scientists working in the same area – have to retrace the passageways provided in the paper and remake the publications and laboratory findings depicted therein in a different way.

Scientific articles are not the only texts that emerge from laboratories. Funding applications constitute an obvious example of another kind of text written in the laboratory and patent applications are becoming increasingly important to universities. These texts are intended to mobilise financial resources rather than scientific credit. Legitimation in the eyes of a broader public also counts as an important resource. This may be achieved through press releases, public presentations and, on a larger scale, through policy-oriented reports on research areas and disciplines. These latter kinds of texts will be discussed only in passing. In this chapter we shall focus on funding applications and patent applications, both because they are important and interesting, and because we shall have occasion to come back to them in Part III of this volume.

When texts go to audiences that are not solely composed of competing colleagues, the forces that operate at the point of reception are different. In order to work upon these forces and create obligatory passageways the text has to recognise the particular requirements and interests of the audience while continuing to function as the emissary of the laboratory. Thus the means by which the actor-world is built up have to be adapted. For one thing, the readers are not in general able to go back to laboratories to check or deconstruct the actor-world they are being invited into. Nor do they want to. Funding agencies and their review panels and assessors judge a funding application in terms of agency goals and criteria. The text thus has to build a bridge between a branch of scientific inquiry on the one hand, and the concerns of the agency on the other. Again, in the drafting of patents, a bridge has to be built between a particular claim and the requirements of patent law and jurisprudence, while policy-oriented reports have to link the concerns of a particular discipline to prevalent social concerns and images of science.

While the scientific article has to work upon the interests of a reader in order to persuade him/her to read the article and pass through the laboratory, the problem confronting those who write the kinds of texts discussed in this chapter is rather different. Readers confronted by a patent or a grant application are normally already interested. The problem of the writer now becomes primarily that of marshalling a set of forces that will withstand sceptical scrutiny. Thus a funding application generally argues that if the funding agency wants to attain some of its objectives, then it has to pass through the research being proposed. A patent assumes that there is an economic interest in its content and sets out both to offer the possibility of exploitation to licencees and to protect its property rights. Texts written for broader publics carefully select material and style in order to reach their audience while convincing it, in the same movement, of the indispensability of science to progress. One could say that in all these cases it is the laboratory that has to displace itself in order to translate successfully.

FUNDING APPLICATIONS

Although the format and rules of funding applications differ in detail between agencies and countries, they nevertheless share certain general features that are connected with the necessity to persuade an external audience both that the proposed research is worthwhile in their terms, and that the applicant is the best person to perform the research. It is thus a matter of interessement (see Chapter 5): the text has to come between the agency and its indifference to the proposed research or its attraction to the proposals of other research groups. At the same time it has to muster and array its forces in such a way that an obligatory passageway is constructed that leads the funding agency through the laboratory of the applicant – to the conclusion that money and other resources should go to him rather than to other researchers in the area.

In order to illustrate this let us take a concrete example which is drawn from the French Action Concertée on fuel cells and the electric vehicle that was discussed in Chapter 2. We are at the stage just before Electricité de France makes its bid for power and presents the actor-world of the future in which urban traffic problems will be solved through the use of electric buses and cars. The original interest in fuel cells is still visible, but this is no longer the exclusive focus of attention. There is also concern to improve the performance of conventional lead accumulators. Now the director of a CNRS Electrolysis Laboratory submits a grant application[1] for one million French francs (anno 1969) to the Délégation Générale à la Recherche Scientifique et Technique (DGRST), the government agency that funds and coordinates the Action Concertée.

The application is for a three-year institutional grant, and sets out a plan of work for the laboratory that is presented as broadly relevant to the Action Concertée. The general structure of this application is one that is found in most such texts. Thus, after the title, the name, and the institutional location of the applicant and a summary of the application, there follow three sections:

(a) a justification for the proposed research and general objectives;
(b) a plan of work, including a list of staff and a budget request;
(c) appendices containing a list of publications and patents that have issued from the laboratory, plus a scientific note on a new development that provides the basis for part of the proposal.

The first section is designed to translate the agency and to link its (presumed) goals and priorities to particular points of passage – the research projects to be undertaken by the Electrolysis Laboratory. It is not, however, a matter of bridging a gap between two given entities. As will be shown below, the research is rearranged to fit with the (supposed) interests of the agency. But the priorities of the DGRST will also be changed when it accepts the application and incorporates the claims being made into its plan.

The second section shows little visible interest-work. The listing of research tasks details and extends the logic of the proposed projects, but there is no further argument about why it is important to do it in this way. The specifications of the personnel of the laboratory and of the budget add to the impression that the second section provides the necessary bureaucratic accountability. It is only because of this that it helps to strengthen the chain of forces: our laboratory is not a phantom, and if your money is spent on us you will get something in return.

The third section is an essential part of the array of forces that is brought to bear on the agency. The fact that it is called 'Appendices' should not mislead the reader for it is necessary to further strengthen the walls of the passage. It is designed to prevent escape by countering the suspicion of those who might feel that the laboratory was not doing anything useful, that its work was scientifically inferior or otherwise unpromising, or that its personnel were inferior and unable to undertake the research in question. Since the text has to stand by itself *all the forces of interessement have to be presented explicitly*.

Let us now turn to a more detailed discussion of the text of the first section of the application. First it should be noted that, as is the case with scientific articles, the forces that are arrayed are heterogeneous in nature. This is particularly obvious here since the policy goals of the agency and the scientific concerns of the laboratory need not mesh. The specific links that have to be made are already revealed in the title: 'Fundamental studies to achieve (1) new batteries and reversible electrochemical cells; (2) fuel cells using hydrocarbons directly.'[2]

In the main body of the text the applicant makes a further attempt to solve his problem: that of showing the technological relevance of his research to the funding agency while simultaneously remaining true to the basic science approach required of those employed by the CNRS. Thus the development of new batteries is not in his brief ('it is not the mission of the Electrolysis Laboratory to develop pilot models, and neither is it competent to do so'), but for precisely this reason it is extremely desirable to collaborate with the laboratories of the CGE (a subcontractor to Electricité de France): 'this is the only way to obtain convincing pilot systems (embodying new battery technologies) within a reasonable time'. The DGRST is assumed to be interested in having working batteries available in the not too distant future, and the applicant creates an organisational and scientific network with himself at a strategic location which will purportedly make this possible.

In fact the problem for the applicant is yet more complicated. Under the influence of EDF, the DGRST has distinguished long-term goals (the construction of vehicles using fuel cells) from short-term possibilities (the design of vehicles using more efficient lead accumulators). The laboratory has thus to insert itself into two different hierarchies of problematisation. It fits itself into the short-term one (short-term transport policy—electric vehicle—rapid development of more effective lead accumulators) by proposing to work on a new type of electrode (with a 'dispersed structure') that may also be used in existing lead accumulators. At the same time it continues to insert itself into the long-term hierarchy of problematisations (long-term transport policy—electric vehicle—ultimate development of fuel cells) by further work on the behaviour of the components of fuel cells. Thus the general objectives of the research are specified in the following way:

a. The continuation of studies on the direct electrochemical oxidation of hydrocarbons on platinum surfaces.
b. Fundamental studies of different systems in which one or both electrodes have a dispersed structure.

Consider the first of these: here the long-term hierarchy of problematisations has been further extended. The problem is now more specific and technical: an analysis of the way in which electrochemical oxidation takes place on platinum surfaces. 'An original method' (the monotubular electrode, see Callon, 1980b) has made it possible for the laboratory to isolate and study most of the phenomena that occur in the electrode. This translation that has been the backbone of the earlier research of the laboratory on fuel cells, now will be further extended. In this way a complete picture of the mechanism of electrochemical oxidation will be obtained. This, then, is the final step of the long-term hierarchy of problematisations – the 'bridge of interests'. It is through the monotubular electrode that those wishing to improve the performance of the fuel cell must pass:

> It is only through the complete scheme of this reaction that one might envisage the means enabling the speed-up of the slow steps in the reaction mechanism, and might define the catalysts best suited to promote these reactions.

A complete series of translations thus transmits the agency from its long-term policy concerns to the reactions of electrodes on platinum surfaces via the laboratory that will undertake the work in question.[3]

The second research objective presents another hierarchy of problematisations. The proposed work, on systems with 'dispersed' electrodes, is a new venture. Earlier in the application it was noted that the laboratory had recently been experimenting with different electrode structures, and in particular dispersed structures. The applicant notes that 'In fact, analysis of the difficulties encountered by industry in producing fuel cells has led us to consider electrode structure from a new angle.' Having positioned himself as a spokesman for industry, he then derives a number of conditions from the requirements of the industrial production of fuel cells, and suggests that dispersed electrodes meet these conditions. While first attempts produced mediocre results and did not meet theoretical predictions, with a special cell the results have been much more promising. This 'early promise' is outlined further, and the text then concludes:

> These results show the interest of continuing such studies, which have been purely exploratory until now, and at the same time of extending the domain of the application of dispersed electrodes to generators that are either new systems that cannot be created in any other way or would otherwise give disappointing results, or to traditional systems (e.g. lead accumulators) whose performance is still susceptible to considerable improvement. It seems clear that such a study – which will, in the long run, contribute considerably to second generation fuel cells – will in the short term lead to interesting and directly exploitable findings in the domain of batteries and reversible cells.

This translation follows the same general pattern as that on the direct oxidation of hydrocarbons. Again, links are made between both the long- and short-term goals of the agency in order to present the proposed research as being of unique relevance to the latter. There is a relationship between the two hierarchies of problematisation: improvement to existing batteries and accumulators arises out of problems encountered with fuel cells and their 'second generation' improvement. However, the work will contribute to the efficacy of first generation accumulators even in the absence of subsequent fuel cells. In fact, in the work plan the main thrust is in research for applications in lead batteries and accumulators. Fuel cells appear only when a study of electrode mechanisms for dispersed electrodes is envisaged.

Overall, however, there is explicit reference neither to the changes in goals of the Action Concertée nor to the scenario specified by EDF. This is partly because the applicant assumes that the background is known to all concerned. It is also, however, a matter of tactics. The director of the laboratory has designed in his proposal an actor-world that may link up with not only the new but also the original goals. To associate the problematisations too closely to the one would risk dissociating it from the other.

In the particular text studied here the difficulties of identifying the goals and priorities of the agency and presenting research as a means to those ends – i.e. creating a map of the world of the future with which the agency is willing to go along – are very visible. However, the hedging of bets is very common in funding applications. To maximise relevance, and therefore the chance of being funded, research is presented as a potential means to as many ends as can plausibly be invoked. Therefore, expected outcomes are not specified in any detail, particularly in more fundamental research. Since the reader does not have to follow any particular one of the specified itineraries, there is a loss of obligatoriness. The overall effect may still be an impression that there will be a general return on the project in question.

Finally, let us consider some of the similarities and differences between funding applications and scientific articles. In the course of the above analysis we have talked of 'hierarchies of problematisation' and the way in which these hierarchies are generated by successive applications and scientific articles. Thus, in Chapter 5, it was argued that the 'funnel of interests' is the characteristic form of translation adopted by the scientific article. Such a funnel selects and chains terms that will (it is hoped) be treated as forces by the readership and hence will lead the latter to an obligatory point of passage. For the scientific article the reader is, of course, the specialist. In the case of funding applications the readership is equally clear. The problem is to assemble and array terms that lead from the goals and priorities of the agency to the particular research being proposed.

It is much more difficult, however, to prevent the sceptic from escaping in the case of the grant application. This is because the doubting grant agency, unlike the doubting scientist, is not being challenged to read all the references cited and replicate the work in question if it wants to challenge the actor-world that is being created. All it has to do is to reject the application by giving a small number of general reasons – it does not have to engage in close argument with the proposer. Accordingly, the author of the grant application reacts by writing in a different manner. As already noted, the tactic of multiplying the relevance of the proposed project may be adopted (with, however, the danger of becoming too vague or grandiose and losing force as a consequence). A complementary tactic arises out of the fact that the agency forms part of a political world and is involved in struggles outside the direct control of any laboratory. The tactic involves (consider Pasteur as discussed by Latour, 1983) presenting research in as 'black boxed' a manner as possible. Thus, instead of asking the reader to consider (for instance) the intricacies of rat yolk sac pinocytosis (see Chapter 5), textual work is directed instead at showing the general relevance of the research programme (and hence its funding) to the agency. In this respect, then, a funding application is like a company share prospectus. Interessement occurs, if it occurs at all, by appealing to past results, anticipated trends and projections of activity.

However, it is important not to exaggerate. It is necessary to supply enough details to make 'early promise' or 'future promise' plausible and to create confidence in the competence of the applicant. The text of the funding application thus has to build a 'bridge of interests' between science and politics, worlds that are generally held to be separate. In the case under discussion here, the applicant generated a document which sought to link the Action Concertée and its interested parties (see Chapter 2) on the one hand, with the scientific research of his laboratory on the other (see also Callon, 1980a, 1980b, 1983).

A corollary to this analysis of the funding application as a 'bridge of interests' is that the reduction of the text to a string of forceful words does not lead to the same sort of result as in the case of scientific articles. In the latter the full force of a text can be captured by a string of words (see Chapters 5 and 7). For funding applications the key-words which represent objects and methods of research may be used to map the world of science that they trace out. However, the words related to the specific structure of means-relevant-to-ends are too local to be used as a basis for maps.[4]

PATENTS

In the texts of patent applications the science is even further displaced than is the case for scientific articles or funding applications. This is because there is a wide range of legal requirements that has to be satisfied, and the applicant has to guard against loopholes and eventual litigation. For this reason, patent applications are often written by specialist patent advisers and not by scientists themselves. Patents are still arrays of forces that are designed to oblige the reader to follow certain itineraries. The itineraries are traced out more explicitly because they are legal documents and not simply texts emanating from a laboratory.

There are three different kinds of readers that have to be led through the text: the officials of the Patent Office, who have to grant the patent; the competitors in the area who are eager for information about new products and processes; and potential licensees who have to be interested in what the patent offers. Thus a patent application is a compromise between different strategies of enrolment. In addition, the application, if granted, becomes a legal document conferring rights on the patentee: it transforms an invention into intellectual property. For this reason, the text as it stands becomes much more important than in the case of scientific articles. Mistakes will often be irreparable. Before considering examples of the strategy and tactics of patent application, we note some general features that stem from the constraints related to the different audiences that have to be managed.

First, the world of science is treated as far as possible as a black box.[5] The patent application is about the uses that a given invention may have, and there is no reason to divulge the know-how to anybody able to read the application in the Patent Office. Patent law, however, requires a description of the invention to guarantee 'enablement': the reader must be able to make and use the claimed invention without undue experimentation. One reason for this clause in patent law is to make it possible for other workers in the area to recognise when they are potentially involved in infringement; the other is the disclosure of the technical advance in return for the protection of intellectual property rights for a limited period of time – in general between fifteen and twenty years. Although patentees may value the patent system in general, in writing their own application they will try to keep the box around their own invention as black as possible while still satisfying the enablement requirement. Penetrability is not in their interests.

The attempted black-boxing of the invention leads us to the second general feature of patent applications: the way in which they map itineraries for the reader to follow. The description of the invention and its uses creates itineraries but adds toll-booths to them. The road to the profitable use of know-how is signposted, but barriers are constructed through patent protection (provided the patent stands up against litigation). Only licensees who pay the toll are able to proceed and allowed to try to reap some of the profits.

Thus the readers are not sucked in by a funnel of interests. Instead, they are challenged by a domain claim in which an interesting area of exploitation is simultaneously indicated and fenced off. Potential licensees have to pass through the toll-booth and pay their dues if they want to gain access to, and obtain a detailed map of, the inside of the black box. Perhaps it would be best to characterise the rhetorical structure of patent applications as a 'fence of interests'. In fact, many patents are taken out not to protect rights to an intellectual product or process, but rather simply to ward off the activities of others in a particular domain.[6]

What are the arrays of forces that are assembled and put together in order to oblige (or forbid) certain passageways? First there is a very common feature derived from the expectation that legal suits will be brought. This is to start with a broad claim that prohibits a wide range of itineraries, but which runs the risk of not being accepted by the Patent Office or of being unable to stand up under later litigation. After this first claim there follow others which are narrower and are accordingly easier to defend, but which bring less profit. To take an example from biotechnology:

> Application N. 008378 of Jackson, Theriault, Sinclair, Fager, and Karwowsky . . .
>
> Claim 2: A process for producing the antibiotic AX-127B-1 which comprises culturing a micro-organism belonging to the species *Micromonospora pilospora* having the ability to produce the antibiotic AX-127B-1 in a nutrient medium including a carbon and nitrogen source and accumulating the antibiotic in said medium.
>
> Claim 3: A process according to claim 2 wherein said micro-organism is selected from the group consisting of *Micromonospora pilospora* NRRL 11415, *Micromonospora pilospora* NRRL 11416, and *Micromonospora pilospora* NRRL 11417, and mutations thereof.

The code numbers after the *M. pilospora* varieties in claim 3 refer to the deposits in culture collections. Such deposits are often made in biotechnology, since the way of isolating ('producing') the micro-organism necessary for a process or product cannot be described exhaustively, and 'enablement' has to be guaranteed by providing a specimen that is publicly accessible.

In this example, claim 2 was rejected by the Patent Office because it held that undue experimentation would be needed for a claim that encompasses unisolated species. However, claim 3 was allowed (Adler, 1984, p. 360). It is not uncommon for scientific articles to present broad claims that are more interesting than the findings on which they are based, but which subsequently have to be retracted as a result of criticism. In this patent application example, however, the breadth of the claim is not related to the interest of the scientific findings but rather to the domain of exploitation. Retraction does not change the science but instead has an influence on the amount of resources that can be mobilised by the text.

Competitors and potential licensees have an active interest in the validity of the patent. If they can show that some passageways are not prohibited (for instance because of insufficient specification), or passage through the licensing toll-booth is not obligatory (for instance because of an earlier publication of the process or product), it will save them money and effort. Accordingly, the attempts to dissociate the forces arrayed in patent applications are often intense, especially when an invention may be very profitable. In order to illustrate this we shall briefly discuss the famous Cohen–Boyer patents that claim to cover the basic recombinant-DNA technology, worth perhaps a billion dollars in licence royalties over its seventeen-year lifetime (even with the liberal licensing policy of their holder, Stanford University). The first patent was applied for in 1974. It was entitled 'Process of Producing Biologically Functional Molecular Chimeras' and described what we would now call recombinant-DNA technology. The patent was issued in 1980 (US Patent 4237 224) after careful consideration by the US Patent Office.

A big problem with this patent is that it is about a process, so infringement of the patent is difficult to trace, and even when its occurrence is suspected, patent law makes it well-nigh impossible to recover damages. Stanford University lawyers had recognised this problem, and therefore filed a second patent, now on a product, the plasmid pSC 101, containing a manipulated DNA. Let us quote a representative claim from this patent application (US Patent Application No. 959 288):[7]

> As a composition of matter, a biologically functional recombinant plasmid capable of selection and replication in a unicellular microorganism cell comprising:
>
> a first DNA segment containing an intact replicon recognized by said cell derived by cleaving a virus or plasmid compatible with said cell at other than the replicon site, which segment is covalently joined *in vitro* at its ends to the complementary end of a second DNA segment foreign to said cell having at least one intact gene, said second DNA segment derived from a source which does not exchange genetic information with said cell.

Since infringement of this claim would be directly detectable, this patent is the really valuable one. If it were granted it would force a large number of actors to pass through Stanford's Patent Department and apply for a licence before they could continue with their work on recombinant-DNA technology.

Because of its potential value, this patent application was screened very carefully by the Patent Office. Though the text of such pending applications is kept secret by the Patent Office, Stanford University decided, for a time, to open its patent file to interested parties. This was, in part, a public relations tactic, but it was also a way of finding out about potential weaknesses at an early stage instead of having to go through costly litigation later on. The invitation to attack the fences of interest and try to bring them down was taken up by at least one recombinant-DNA technology user who had not arranged for a licence with Stanford University. A patent attorney at Exxon published a review article, 'Patenting the Results of Genetic Engineering Research: an Overview', in an academic volume. Knowing that the Patent Examiner in charge of the Cohen–Boyer patent could read the paper, he used footnotes to present information that might lead to rejection of the patent application.

One potential objection arose from the fact that a publication about the plasmid had appeared in the *New Scientist* more than a year (i.e. more than the 'period of grace' allowed by US patent law) before the patent application was filed. To the relief of science journalists and others who were expecting molecular biologists to become even more close-mouthed than had previously been the case, this objection was waived by the Patent Office. The second potential objection hinged on a technical point: the application should describe the 'best mode', but Cohen had noted, in a later publication, both that the plasmid as described would not do what it was claimed it could do, and that he had not, at that time, deposited the plasmid.

Consider the way in which the Exxon patent attorney (Halluin) attempted to dissociate Cohen and Boyer's array of forces at this particular point. After making a general remark in the main text to the effect that developments may prove an original procedure to be inoperative in producing the desired plasmid, he footnotes the sentence with a lengthy discussion of the Cohen–Boyer patent. Part of the text, with the comments of Goldstein (1982) that highlight some features, is given below:

> It is interesting that such a revised interpretation has occurred with respect to the origin of plasmid pSC 101 [citing a Cohen and Chang 1977 article in *J. Bacteriology*]. In this article, Cohen *et al.* state that

> the original belief that pSC 101 was formed by recircularization of a fragment of R6–5 DNA was incorrect. This observation is particularly interesting [note how 'interesting' makes its appearance once again: a lawyer's way of saying, 'Hey! Something funny is going on!'] in view of Dr Cohen's statement in his article in *Scientific American* (Cohen, 1975) which reads: 'For a time after our initial experiments the pSC 101 plasmid was the only vehicle known to be suitable for cloning DNA in *E. coli.*' . . . It is noteworthy [!] that Dr Cohen did not deposit pSC 101 with the ATCC (37032) until June 25, 1981 (pers.comm. B.A. Brandon, of the ATCC).

Although Halluin purportedly only makes interesting observations, he is in fact saying that the patent may not enable the best mode. On this and other grounds the US Patent Office withdrew the case in 1982. Stanford University lodged a rebuttal, and most objections were met or judged insufficient. The Patent Office granted the second patent in September, 1984. Even so, the reopening of the patent file has revealed that the 1977 paper by Cohen only became known to the Patent Office as a result of Halluin's article. This was in violation of proper procedure and, despite the granting of the patent, suggests that the latter may prove difficult to enforce in practice.

What concerns us here is how patent law and jurisprudence set the scene for interessement and counter-interessement: enablement, publication problems, 'best mode' and so on, these lay out a scenery of forces that are used by the different parties to erect fences or to take down the fences that constitute their opponents' claims. In the particular example discussed here, Exxon attorney Halluin, in trying to influence decisions of the Patent Office, not only shows how the forces of the world of patent law are borrowed to change passageways – if Halluin's views are upheld by the Patent Office, Exxon need not pass through Stanford University to use a convenient plasmid. It also shows how scientific developments, with their own dynamics, interact with the world of patents. Thus, even if science is black-boxed in a patent application, the box may be prised open, and the content of scientific articles may be used by patent attorneys to bring about their counter-interessement.[8]

In conclusion we note that the texts of patents are structured according to patent law, jurisprudence and litigation, but borrow some of their force from science, and the invention (assumed to be science-based in our discussion) has to be detailed sufficiently to meet the requirement of enablement. This part of a patent text can be reduced to a string of key-words, and form the basis of scientometric analysis (see Chapter 10). However, as with funding applications, such a string of key-words does not cover the full force of the patent text. To trace the 'fencing of interests' a separate analysis of each patent and its context is necessary.

CONCLUSION

The question as to how texts mobilise resources cannot be answered in a general way. Hower, the analysis offered in this chapter, although preliminary, already reveals some interesting features and suggests that the approach is fruitful.

It is striking that science cannot normally just claim resources and expect to receive them. Rather, it has to orient to given arrays of forces – those of funding agencies or patent law – and build obligatory passageways by modulating them. This is clear in the means–ends rhetoric of funding applications. Similar rhetoric can be found in a more diffuse and indirect form in prospective reports, where the general utility and/or fundamental importance of whole disciplines is argued. Because public legitimation and a favourable climate in government agencies is sought, the argument has to accommodate itself to prevailing opinion and images of science – a diffuse but very real requirement. Since the audience is undefined, the rhetoric is not very tight or directed. One may speculate, however, that the demands of science policy and the need for legitimation will lead to further articulation and specification.

Patent applications have a different shape because the resources they mobilise benefit science only indirectly. They have to follow legal requirements in order to realise property rights based on science. The 'fences of interest' that they erect lead to specific and explicit strategies for the construction and deconstruction of arrays of force. In the course of these struggles scientific findings may become involved.

An interesting aspect of patent applications is that when the patent is finally issued the text, as such, certifies legal rights. Its array of forces is declared durable by the Patent Office and stands between the new process or product and potential users. Therefore the details of the text become much more important, because irreparable, than is the case for scientific articles. Legal battles may now be fought about interpretations of words or sentences.

In general, the texts sent out by laboratories to external audiences have to stand by themselves. The particular power of scientific articles, their links to the laboratory and to other articles, can be exerted only sparingly. Policy-makers, patent officials, grant-awarding bodies, these work primarily with the text that is submitted to them. Deconstruction of the assembled forces follows political, legal, or bureaucratic lines. The scientific component of the text has to meet minimal requirements but is not a direct object of criticism.

Although we have emphasised that science has no direct power and has to insert itself in an external world, it would be wrong to picture it as a beggar or a supplicant, satisfied with the crumbs that fall from the tables of princes. The texts are able to present science as an obligatory passageway for government and industrial firms. The 'palace' and the 'firm' are interested in the 'laboratory'. There is, then, something of a paradox here: science is inevitable and indispensable in contemporary societies, but when it mobilises resources it cannot use its power to command them.

In order to see how power struggles between the 'palace', the 'firm' and the 'laboratory' are conducted it is not enough to look at the structure of texts. In our study of the workings of laboratories we argued that it was first necessary to follow the actors, and only subsequently consider the working of texts as one of the instruments used by the actors in their struggles. The same approach will be necessary for what used to be called the external relations of science: we are in need of ethnographies about funding agencies, patent offices and committees on the public aspects of science. Only then will we really begin to understand how the texts sent to them by scientists are able to mobilise resources.

NOTES

1. 'Recherches fondamentales en vue de la réalisation; premièrement de nouvaux accumulateur et piles réversibles, deuxièmement de piles à combustion directe des hydrocarbures' (Laboratoire d'Electrolyse du CNRS, 1969).
2. All translations from the French original are by Arie Rip.
3. But, lest the DGRST escape, a secondary interessement is advanced: if one is interested in widely usable fuel cells then the direct oxidation of hydrocarbons constitutes the only really acceptable solution. Alternative developments that lead to application in a nearer future need extra parts and will therefore not be widely used. For good measure it is noted that 'The same assessment has been made by experts of the US Army'.
4. Unless, of course, one were particularly interested in the way the world of an Action Concertée was mapped in funding applications, and also prepared to draw up a special set of key-words and a coding procedure to this end.
5. Many patents are not science-based, but we limit our discussion to those that are.

6. In some countries this strategy is counteracted by requiring actual application of the invention for the patent to be upheld.
7. The patent claims are quoted after Goldstein (1982).
8. Earlier in history, the interaction between scientific developments and patent cases was much more direct as, for instance, in the early dyestuff chemistry studied by Van den Belt and Rip (1984). Advances in science could even lead to modifications in patent law (routinisation of innovation in the case of azo dyes), as is now happening again in at least some countries for the patenting of engineered organisms, see Adler (1984) and Huni and Buss (1982).

Part III
Mapping Science and Technology

7 Qualitative Scientometrics

MICHEL CALLON, JOHN LAW and ARIE RIP

INTRODUCTION

Scientific and technical research is in a state of continuous flux. Sectors once considered to be strategic become obsolete, while new and promising fields appear. There are bottlenecks, points of obligatory passage, obsolescent sectors, areas of growth and areas of fusion between previously separate domains of research, while external factors are of frequent importance. The importance of the analysis of the dynamics of science and technology is scarcely in doubt and it has, accordingly, attracted much interest.

A qualitative concern with scientific change is to be found in a range of disciplines. It has been a central concern of philosophers (see Popper, 1959, Lakatos, 1970, Toulmin, 1972 and Hesse, 1974), and sociologists of science have displayed a similar concern in their many case studies of scientific controversies (e.g. Collins, 1975, MacKenzie, 1978), specialty growth (e.g. Mullins, 1972) and day-to-day work in scientific laboratories (e.g. Latour and Woolgar, 1979). Such an interest also underlies the work of many historians of science (e.g. Kuhn, 1970, Shapin, 1979 and Thackray, 1972), and such writers on science policy and the political direction of science as Weingart (1982) and members of the Starnberg school (Rip, 1981; Böhme, van den Daele, Hohlfeld, Krohn and Schäfer, 1983). Though these many and varied writers have advocated a wide variety of theoretical perspectives they all have one thing in common: they do not normally make use of quantitative indicators or utilise algorithms in order to handle aggregated data.

Quantitativists have worked in a quite different way. Using large computerised databases they count publications, citations and patents. Though these counting procedures have been used for a range of different purposes, this work also results in a representation of the state of development or evolution of a given domain of science or technology. Thus the development of co-citation clustering techniques pioneered by Small and Garfield very clearly reveals the interest of quantitativists in the dynamics of science and technology.[1] As a general rule, however, such writers have displayed greater interst in the development of quantitative methods than in the elaboration of the theoretical significance of methods or findings.

Qualitativists and quantitativists thus share a common interest in the dynamics of science. However, it is the argument of this book that if further academic or policy-relevant progress is to be made it will be necessary to build on this convergence. First, the reluctance of the qualitativists to collect aggregated material appropriate for statistical analysis must be overcome (see Edge, 1979). Second, the kind of material collected and the uses to which it is put must contribute to a theoretically defensible conception of science and technology.

In Part II of this volume a new theory of the dynamics of science and technology was spelled out. In the second section of this chapter we highlight aspects of this argument and outline the way in which a quantitative method consistent with this theoretical approach may be derived. In the third section we present and discuss the actual tool of co-word analysis, and in the final section we describe typical results that may be obtained using this method.

THE DYNAMICS OF SCIENTIFIC AND TECHNICAL DEVELOPMENT

A Theory of Scientific and Technical Change

Initially it is necessary to avoid two kinds of reductionism: first, that which assumes that science and technology develop in terms of some kind of internal logic; and second, that which assumes that these can be explained in terms of external class, economic or ideological factors. Scientific and technological action is an emergent phenomenon, explicable neither by nature or reason on the one hand, nor social influence on the other. Rather the scientist or the technologist should be seen as an entrepreneur whose goals and resources are both defined in a process of interaction with other actors. For Electricité de France electrons are neither more nor less necessary than Renault, customers, municipalities or the Ministry for the Quality of Life. For the Chinatown biochemists, rats are neither more nor less important than ^{125}I radio-labelled PVP, the laboratory apparatus, or the scientists enlisted as literature references. There are no necessary links between such heterogeneous entities. The construction of scientific facts and technical artefacts involves the creation and consolidation of such links into an array that is strong and durable. Accordingly, the histories of these more or less durable associations provide the key to an analysis of development in science and technology.

The concept of the 'actor-network' was introduced to describe such associations. Actor-networks grow by enrolling other entities. They do this by 'translating' the projects and purposes of these entities, and establishing themselves as the spokesmen of those who are being associated. To translate is not, however, just to speak for others. It is also to displace them and let them work for your projects rather than their own. To manage this, the translator must make the displacement either desirable or unavoidable from the standpoint of the entity being enrolled. If Renault does not want to disappear, swept away by a post-industrial society in which there is no room for vehicles with internal combustion engines, and if the fuel cell does not want to end up in a dark storage room in a laboratory, then they have to enrol in the project of the Electricité de France and make an obligatory passage through the electric vehicle. The analysis of scientific and technological development depends upon the description of such translating actor-worlds, their development, their struggles with each other, and the compromises which may be forced on them. We have to find ways to identify and trace the translations. (For a further discussion of translation see Callon, 1980b, 1985).

Though there are many strategies for translation, they all redefine the identities of entities by drawing them from one context into new roles in the world of the actor-network. A common mechanism for deflecting and enrolling entities is that of positing an equivalence between two problems (see Callon, 1975; Callon, Courtial, Turner and Bauin, 1983). Electricité de France suggests that Renault can only solve its problem of survival in a post-industrial society by *first* contributing to the solution of the problem of how to produce an electric vehicle. In the process of making equivalent, texts are of particular strategic importance. Are you interested in chemotherapy? Then you must, of necessity, be interested in the behaviour of cells. And if you accept that these are equivalent, there are further displacements on the way which guide you to pinocytosis and the polymers that influence this phenomenon. Once you reach this point the work of the text is nearly done: you will inevitably be attracted to the study of the effects of DIVEMA on cell membranes. The juxtaposition of equivalences has brought you to new, but vitally important, terrain. You have been interested and canalised (Law, 1983).

Thus the words of a text are operators, strung together to impose a series of displacements on the reader, where each displacement derives from a postulated equivalence between problems. Though translation does not work through texts alone, the latter nevertheless reveal a great deal about translation in science and technology. First, texts are ubiquitous in research, emissaries of particular importance to the actor-networks that send them out. Second, they are constructed in order to exert as much force on the readers as possible (see Chapters 4 and 5). Third, the way in which they attempt to exert this force is somewhat transparent, not because a literal reading of any text is possible, but because they work by the visible marshalling of their forces. Thus the translations or postulated equivalences between problems appear in scientific texts as associations of words which signal and fix the interests-cum-problems which they displace and link together. 'Tumour', 'therapy', 'pinocytosis' and 'DIVEMA': these four words associated in a scientific article are like a *pointille* picture of the actor-world. For the purposes of analysis, then, an article may be reduced to a network of powerful words which can be represented as a simple graph, as in Figure 7.1

Each point represents a word, while a line indicates that the words linked in this way co-occur in the article. Thus the graph that summarises the article can be read, like the original article, as a structured network of claimed equivalences between problems which may be of an entirely heterogeneous nature – in this example a

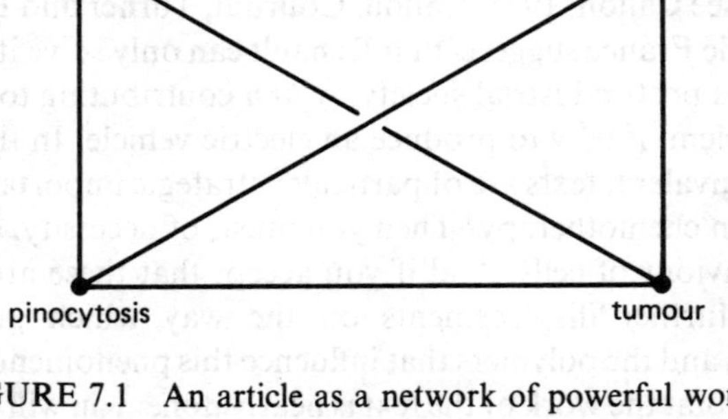

FIGURE 7.1 An article as a network of powerful words

molecule, a biological effect, a method of treatment and a health scourge. Such claimed equivalences between what may be quite different elements constitute the authors' strategy of translation. Thus the actor-network of which they present themselves as spokesmen, is immediately described through the world of equivalences between problems that it builds. There is no need to define actors separately from their products.

Interaction Between Actor-Networks: A Problem of Method

By focusing on the nature of a single scientific text and its transformation to a graph we have taken the first step towards tracing the dynamics of science and technology. But as actor-networks send out texts and hope to grow by translating the forces cited in their texts they inevitably come up against other actor-networks. The dynamics of the production of knowledge arise from these confrontations, competitions, conflicts and their eventual (more or less temporary) resolution. The problem then becomes: how can we distinguish between powerful texts which allow their actor-networks to grow and those that are weak and vanish into oblivion? For victories, defeats, growth and disappearance all arise from interaction *between* the texts of actor-networks. Part of the problem, then, is to study texts at a higher level of aggregation.

The success of an attempt at translation depends upon the extent to which other texts (articles, reports, patents) take up and make use of its structure of problems. Where a given translation or association is widely adopted then its strength is assured. However, when a researcher is alone in proposing a translation and others are not enrolled, the actor-network in question is weak, and the associations it proposes lack strength. A count of the number of times that an association is repeated is a check on the solidity of the actor-world and its associations. Obviously, to do this, many texts have to be examined and qualitative studies are insufficient. It would take too much time and effort to multiply detailed case studies and for practical reasons alone one needs to supplement these with a routinised, quantified approach. The refusal of those who adhere rigidly to qualitative studies to count thus makes it extremely difficult to trace the successes and failures of attempts at translation. How can one find out whether the behaviour of electrons on platinum surfaces has to be studied if urban traffic problems are to be improved? One possible way is to measure the number of times such an equivalence appears in the total body of texts in which it might occur. Thus the 'co-word' approach does not indulge in counting for counting's sake. Our problem, that of tracing the dynamics of science and technology, demands the use of quantitative methods, but these are best seen as the pursuit of the qualitative by other means.

The importance of the quantitative is further highlighted by the fact that it is only in certain cases – such as those in Part II concerning the electric vehicle and DIVEMA – that direct links are made in particular texts between laboratory work on the one hand and social problems and interests on the other. Normally, though these exist, such links are invisible in the maze of translations that arise from the interaction between actor-networks. Under these circumstances it becomes difficult if not impossible to trace the relationships between the world of research and the world of industry or consumers. This explains why qualitativists often jump from detailed analyses of scientific controversy to general explanations posed in terms of social interests: they are unable to make the connection in a more detailed and less perilous manner. By contrast, the co-word approach, by summarising articles in terms of forceful words and counting occurrences and co-occurrences to trace developments at an aggregate level, not only allows successful translations to be traced and distinguished from those that quickly disappear; it also makes it possible to uncover the many direct and indirect links that exist *between* translations whether or not these lead rapidly to social problems and interests.

Networks of Problematisation

A scientific article translates by positing a small number of problematisations – i.e. equivalences between a problem or problems that interest a reader and the problem that the authors want the reader to pass through. Consider a single problematisation: $P_1 \sim P_2$. Here, if it works at all, the relationship of equivalence operates in both directions, even though the authors may be primarily interested in claiming that the resolution of P_2 has to pass through the solution of P_1. As they work on their problematisation the authors are able to utilise existing, more or less accepted, equivalences (for instance $P_1 \sim P_3$ and $P_2 \sim P_4$). These will appear in their text and will be depicted as lines connecting the corresponding words in the graph. In other articles by different authors similar equivalences may be used, but additional ones (for instance $P_1 \sim P_7$, $P_4 \sim P_5$ and $P_3 \sim P_5$) may also be cited. The situation begins to become more complicated. Figure 7.2 visualises the network of linkages around a single paper.

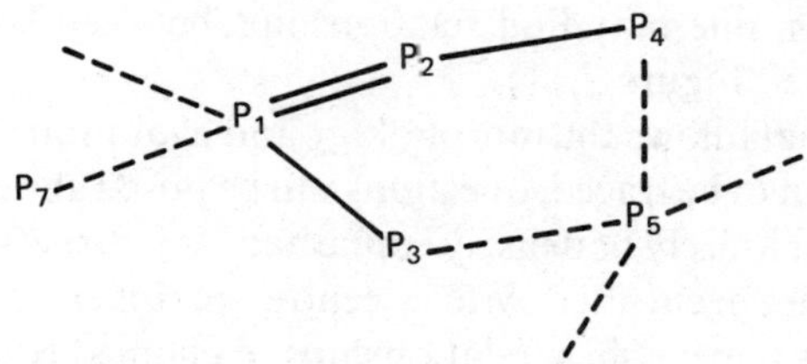

FIGURE 7.2 The network of problematisations around a single paper

The overall linking of problems depends on the crisscrossing of a large number of separate strategies of translation. Thus one may visualise the literature in an area of science as the embodiment of a network of problematisations as shown in Figure 7.3. Such a network represents the repertoire of translations available in a given area at a given moment, and allows the authors of texts to pass from one problem to another, often with little effort.[2]

Obviously such a network of problematisations is never permanent. Its stability cannot be more than that of a shifting equilibrium between the forces working to generate translations. Each scientific text, while drawing on the available network of problematisations, also contributes to it and may change its structure.

Our approach – that of adding up the co-occurrences of forceful words in texts – highlights the networks of problematisations and enables us to study their morphology in detail. Thus Figure 7.2 suggests that one may discover linkages in the network which do not exist explicitly within any given text. P_1 is linked to P_5 through P_3, but since P_4 and P_5 are linked and P_4 is taken as equivalent to P_2, there is an additional route from P_1 to P_5 through P_2 and P_4. As a result of such

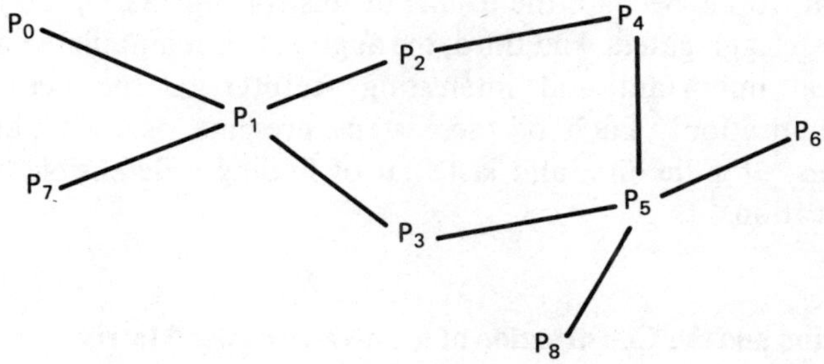

FIGURE 7.3 The network of problematisations in an area of science

indirect linkages, one may find relationships between 'spawning' and 'cage culture' (see Figure 7.11).

With a tool that allows the morphology and evolution of networks of problematisation to be traced, questions can be posed about the shape of the network. Is it loosely or densely connected? Is it dispersed with many small nodes or concentrated with a centre–periphery structure? Additional features appear since relationships of claimed equivalence may in fact be asymmetrical. Thus problem P_2 may be generally accepted and available without effort, while problem P_1 is still struggling for recognition by allying itself with P_2. Such an asymmetry will be reflected in a higher frequency of occurrence of the word representing P_2 than that representing P_1. An analysis of such differences in frequency reveals hierarchies and allows their evolution to be traced. Problems of minor importance may be identified and their role in restructuring the network if they grow in importance may be studied. In addition, the actors (researchers, agencies, enterprises) responsible for the formulation and interrelation of the different problems may be identified while those that have the greatest influence on changes in the network may be located. Finally, it becomes possible to determine both those actors responsible for the minor problems that struggle to subvert the hierarchy of network structure and those, by contrast, that tend to support the status quo. The approach, however, assumes that a concrete tool is available which can generate maps of problematic networks. It is to a description of such a method that we now turn.

THE METHOD OF CO-WORD ANALYSIS

There are three stages in the identification of networks of problematisations. First, individual texts are transformed into a string of key-words or signal words. Second, the results of this transformation on a set of papers are aggregated. And third, the aggregate is manipulated in order to reveal important and interesting features of the network of problematisations. Each of these steps presents its own particular problems – for the first and last that of finding a defensible form of simplification.

Indexation and the Construction of a Co-Occurrence Matrix

In principle one might do content analysis of all texts and use a set of key-words derived from the texts in question and adapted to the specific purposes of the analysis (Rip and Courtial, 1984). This, however, would be very expensive. In practice we use the summaries produced by services for scientific abstraction and documentation. In particular, we use the French indexing service, the CDST, which publishes summaries of scientific articles, patents, reports and (mainly French) theses that have been indexed in terms of key-words drawn from a recognised lexicon. Obviously the intervention of indexers introduces uncertainties into the relationship between the key-words and the text itself. As we shall show in Chapter 11, the key-words assigned in this way represent a workable version of the problems-cum-interests in the text – that they constitute an acceptable rough and ready way of mapping the structure of the actor-world that will be imposed if the text is successful. What, then, is the basis of this assumption?

No literal reading of a text is possible (see, *inter alia* Garfinkel, 1967) and the impact of a document depends upon the readings of it given by others and the extent to which these in turn use its translations in their own work. However, a professional indexer is also an agent of translation. Although his or her task is to be true to the contents of the article, she/he introduces key-words with the users of the documentation service in mind. Thus articles on both arteriosclerosis and constipation are indexed by a range of specific words but the term 'dietary fibre' may be added to these if it is felt that the articles in question belong to a single category for certain users.[3] Since scientists and documentalists continue to use these documentation services, one may conclude that indexing does not distort too much, at least from their standpoint. The main effect is rather addition to the network. By contrast, what may not be detected by the documentation service are new categories used by a handful of authors. Even in this case, however, it sometimes happens that a new combination of existing key-words becomes visible which is only later replaced by the new concept. As we shall show below, there are ways of making such 'growth points' visible in the aggregation of key-words. The claim that documentalists add to rather than distort networks of problematisations is additionally supported by studies in which the results of co-word analysis have been compared with those of qualitative investigation. The results of co-word investigations, undertaken by investigators with no substantive knowledge of the area in question, have been taken by specialists in certain areas to be a broadly accurate depiction of their field. This finding is discussed more fully in Chapter 12.

The second step, the aggregation of the results for individual articles, is relatively straightforward. An appropriate set of criteria is used to establish a file of articles (see Chapter 11), each of which has its own bibliometric data and string of key-words. A lexicon is formed of all key-words occurring in the set. Two key-words co-occur in the set if there is at least one document which has both M_i and M_j in its string of key-words. The strength of the correspondence is expressed as c_{ij}, the number of documents in which the couple (M_i, M_j) appears. The co-occurrence values c_{ij} can be arranged in a symmetrical matrix. It is convenient to put the diagonal elements c_{ii} equal to c_i, the frequency of occurrence of key-word M_i in the set of articles. In this way a co-occurrence matrix is constructed that contains information about the aggregated network of problematisations for the domain in question.

The Indices of Inclusion and Proximity

The final step in the analysis is to highlight the interesting features of the co-occurrence matrix. Since different questions may be asked about the network of problematisations the co-occurrence matrix is subjected to a variety of different operations. Here we shall concentrate on two of these – hierarchies, and minor but potential 'growth' areas. In this way we are able to depict the 'essential tension' or dialectic between mainstream orthodoxy and marginal heterodoxy.

The hierarchy of problematisations can be highlighted by calculating an inclusion index. Key-words which occur more than three times in the file and co-occur above a certain threshold value of the inclusion index appear as points on a graph, connected to the other key words with which they co-occur significantly. The technique is straightforward, though interpretation depends upon an understanding of hierarchisation as a property of networks of problematisations. The inclusion index I_{ij} is defined as:

$$I_{ij} = \frac{c_{ij}}{\min(c_i, c_j)}$$

and can be interpreted as a conditional probability. If $c_i > c_j$, the inclusion index measures the probability of finding M_i in an article given that M_j is one of its key-words. To take an extreme example, if $I_{ij} = 1$, the occurrence of M_j implies that M_i will also occur in each article. In such a case M_j is fully included in M_i, the more general key-word. In terms of the theory outlined in the second section of this chapter, the equivalence from M_j to M_i is accepted as automatic and no effort is necessary for the translation. Note that the opposite is not true: if M_i occurs more frequently than M_j, the translation from M_i to M_j requires many

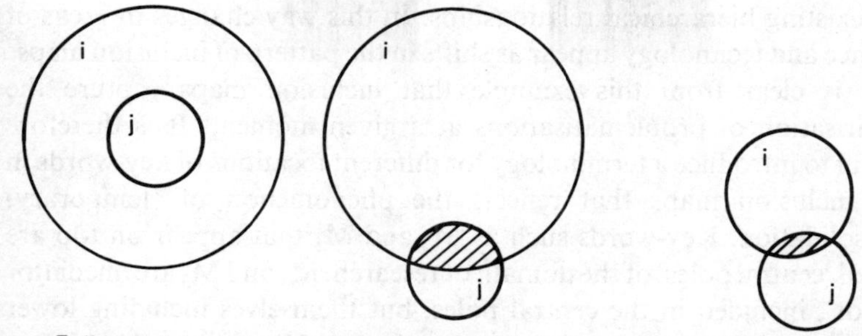

Full inclusion of M_j in M_i — 50% inclusion of M_j in M_i — Low inclusion of M_j in small M_i

FIGURE 7.4 Visual representation of inclusion index

restrictions and specifications. (For a visual representation of three cases see Figure 7.4.)

Since I_{ij} is not symmetrical, the link between points i and j on the graph can be drawn as an arrow from j to i, and an oriented graph results. If we introduce a vertical scale into the oriented graph (usually the frequency of occurrence of the key-words as the ordinate), inclusion maps are thereby obtained that reflect the hierarchical relations between key-words and present them in an easily recognisable way, as in Figure 7.5.

This should be compared with the network of problematisations depicted in Figure 7.3. The hierarchical relationship between P_2 and P_1 is reflected in the inclusion of M_1 in M_2. The other key-words have been assigned their positions in order to highlight a feature that is often found in inclusion maps. There is a straightforward relation between M_1 and M_5, mediated by M_3. There is, however, also an indirect route, passing through M_2 and the low-frequency key-word M_4. If the latter key-word were to become more important it would rise on the map, and transform

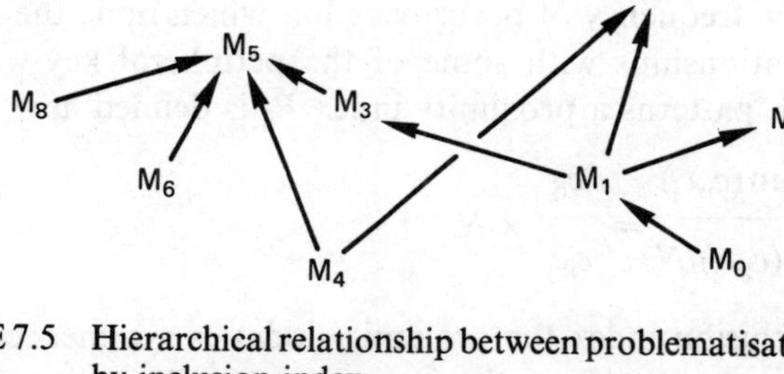

FIGURE 7.5 Hierarchical relationship between problematisations as depicted by inclusion index

the existing hierarchical relationships. In this way changes in areas of science and technology appear as shifts in the pattern of inclusion maps.

It is clear from this example that inclusion maps capture the polarisation of problematisations at a given moment. It is therefore useful to introduce a terminology for different locations of key-words in the inclusion maps that reflects the phenomenon of (temporary) consolidation. Key-words such as M_2 and M_5 that appear on top are called 'central poles' of the domain of research. M_1 and M_3 are 'mediator words', included in the central poles, but themselves including lower 'peripheral issues' – those that are marginal with respect to the domain of research in question. (It is important to note that these key-words might be found in quite different hierarchical positions if another domain were to be studied.) These hierarchical positions can be used to structure graphs vertically in place of the ranking in terms of frequency mentioned above, and in later chapters this is the convention that is generally adopted.[4]

A final convention also increases legibility. The peripheral key-words are arranged horizontally in such a way that the number of intersections between connecting arrows is minimised. It should be emphasised, however, that metric distances on the maps are no indication of the intensity of relationships. The strength of inclusion between key-words is determined by their co-occurrence.

Inclusion maps are designed to highlight the central poles in a domain and depict their relations with key-words that occur less frequently. To trace the networks of problematisations based on significant equivalences that occur only in a small part of the domain it is, however, necessary to operate on the co-occurrence matrix in another way. The following argument can be made. Even if the inclusion index has a low value – i.e. the probability of finding M_i in an article given that M_j occurs in it is low, this probability might still be significantly greater than the unconditional probability of finding M_i in any one of the N articles in the domain. Such a situation will obtain for mediator key-words which have a relatively low frequency of occurrence but which none the less have significant relationships with some of the peripheral key-words. To bring out such patterns a proximity index P_{ij} is defined as:

$$P_{ij} = \frac{c_{ij}/\min(c_i,c_j)}{\max(c_i,c_j)/N} = \frac{c_{ij}}{c_i c_j} \times N$$

Unlike the inclusion index the proximity index is symmetrical, and a proximity map consists of a collection of larger and smaller clusters of

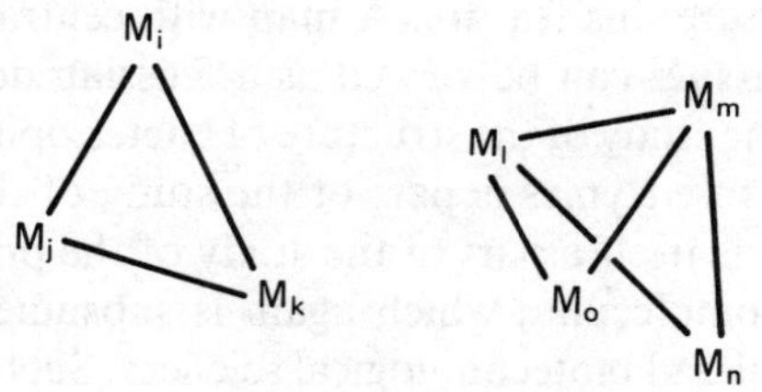

FIGURE 7.6 Clusters of problematisations as generated by the proximity index

key-words which, in their basic form, usually involve the co-occurrences of three or four words (see Figure 7.6).

If the threshold value is lowered then more proximity connections between key-words appear on the map and, in the end, the mediators and central poles found on inclusion maps reappear. This is one way in which relationships between 'minor' issues and the pattern of consolidation in a domain can be studied, though clusters may exist even where the domain has no overall structure as this is represented by the central poles of inclusion maps. (For a visual representation of high and low values of the proximity index see Figure 7.7.)

To summarise, not all domains reveal a structure of hierarchical relations as depicted by the inclusion index. Co-word analysis thus makes possible the empirical study of the degree of consolidation in a domain – the extent to which it is internally structured by 'entrenched concepts' (to use Mary Hesse's (1974) term) – though most domains display some degree of hierarchisation. Simplification or 'black-boxing' are general features of science and technology, and the inclusion index is

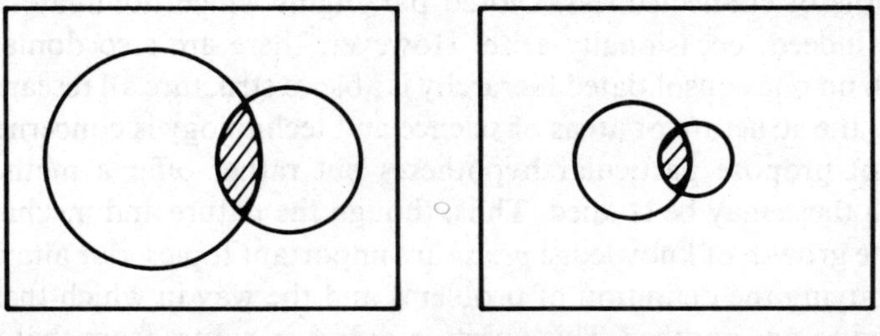

Low proximity (10 co-occurrences) High proximity (10 co-occurrences)

FIGURE 7.7 Visual representation of proximity index

designed to capture this feature. A map with central poles, mediators and peripheral issues can be viewed as a Russian doll, a set of nested black-boxes.[5] The study of the structure of bacteriophage RNA with the help of restriction enzymes is part of the study of the organisation of genomes, which is itself a part of the study of the primary structure of biological macromolecules, which again is subsumed under a central pole of biological and biotechnological sciences. Such a structure can be subverted but there are forces that tend to maintain and consolidate it.

The central poles, for instance, general words that denote themes of interest in the life sciences and their technologies, relate to the work of thousands of researchers and engineers, as well as to funding agencies. The relationship between a laboratory and the science policy of a ministry can thus be traced using inclusion maps. This provides an answer to the fundamental question of how sciences and technologies are enclosed in black boxes without being isolated from the rest of society. The mystery of the links between dollars and markets on the one hand, and spectroscopic analysis on the other, is dissolved when it is possible to trace chains of problematisations through hierarchies of black-boxed key-words on inclusion maps.

The importance of the proximity maps lies in the way in which they show up connections between minor problematisations that would otherwise remain hidden behind the 'entrenched concepts' of the central poles. Though many remain for ever minor, such peripheral issues may profit from their very obscurity and gather strength in an attempt to restructure the network of problematisations. This dialectic between the entrenched on the one hand and the peripheral and possibly innovative on the other, in some ways resembles Kuhn's (1970) distinction between paradigms and the anomalies that may lead to a revolution. For us, however, the difference is one of quantity rather than quality. There is nothing other than networks of problematisations of different sizes and patterns of connectedness. Closed paradigms which dominate a field may, indeed, occasionally arise. However, there are also domains in which no one consolidated hierarchy is able to structure all research. So far as the structure of areas of science and technology is concerned, we do not propose particular hypotheses but rather offer a method by which these may be studied. Thus, though the nature and mechanisms for the growth of knowledge *per se* are important topics, our aim is that of studying the definition of problems and the way in which these are related to one another. This question differs in nature from that of the growth of knowledge, but it is one which the maps of inclusion and proximity are well able to answer.

EXAMPLES OF CO-WORD MAPS AND THEIR INTERPRETATION

In this section we wish to give some idea of the range of findings that may be obtained when co-word analysis is used in earnest on scientific databases. We also want to underline the fact that co-word maps are never transparent in meaning. Their use requires an empirically sensitive interpretation that rests upon the notions of problematisation, hierarchisation and simplification discussed above. In other words quantification is not an end in itself, but rather a method, albeit a vital one, for the pursuit of the qualitative by other means.

As a first example we present the Leximappe (Registered Trademark) of the domain of 'dietary fibre' research for the period 1973–6 which is based on the key-word coded articles in the Pascal database of the CDST, with a threshold of $I_0 = 0.05$ (see Figure 7.8).[6] In fact the label, 'dietary fibre' is an anachronism, because only after 1976 does a hierarchy appear with the term 'dietary fibre' located in a prominent position (see below).

This map displays a multi-polar structure. A first group of problematisations is headed by the key-word 'alimentation' (food substances), a second by the key-word 'colon' (colon), a third by 'régime enrichi' (enriched diet) and a fourth by 'traitement' (treatment). By themselves these key-words mean little. In order to interpret them it is necessary to study the mediating and peripheral key-words which they aggregate. Thus it appears, for instance, that 'colon' brings together problematisations about the dietary treatment of various diseases of the colon. It is important to emphasise that inclusion maps do not present research problems as such, but rather the hierarchical structure of their successive simplifications. It is easy to imagine concrete situations in which simplifications might occur. Consider, for instance, a conversation between the head of a scientific agency and one of his scientific directors, the former enquiring of the latter about the main issues in research on dietary fibre. The scientific director, knowing that he will be asked for further details, mentions only the four poles indicated above. To fill out this picture for the annual report he will go to his team managers and they will provide the mediating issues subsumed under these poles. The simplifications are thus successively 'de-simplified' and the graph is decomposed, if necessary until the peripheral issues appear.

This little exercise should not be taken to imply that there is necessarily one scientific agency that covers all the problematisations that appear in the inclusion map of the domain of dietary fibre. Rather,

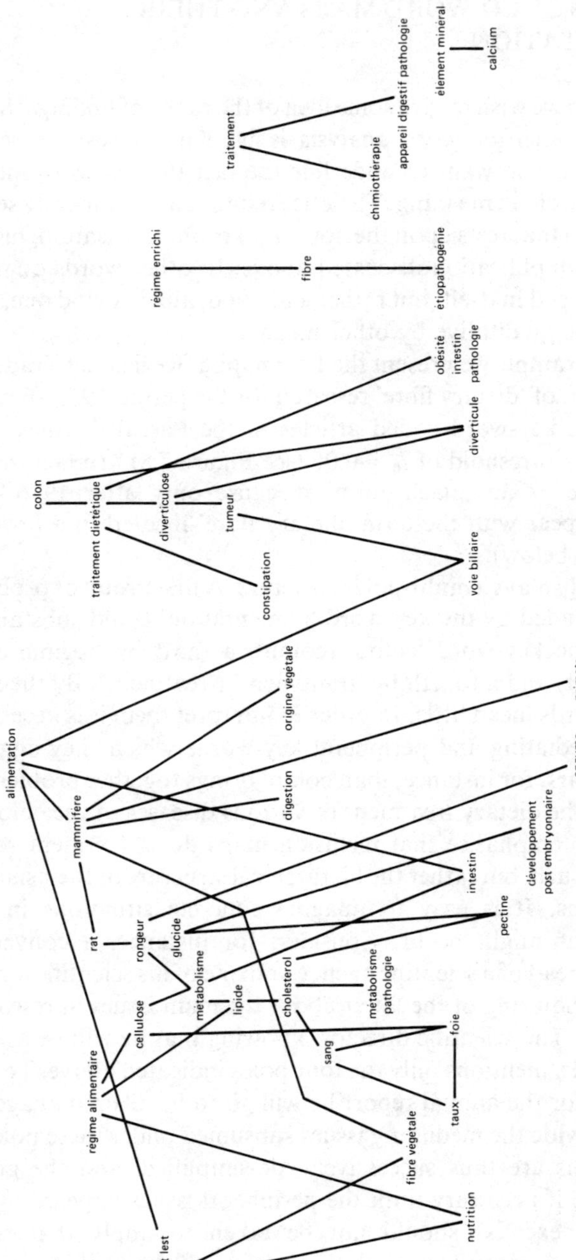

FIGURE 7.8 Inclusion map for dietary fibre research, 1973–6

it displays the *logic* of the type of questions to which inclusion maps provide answers. There may well be no organisation that is interested in dietary fibre as such. But the different actors working on problematisations in this domain are able to define and control themselves and each other because they have a repertoire of problematisations available to which they contribute and by which they are constrained. It is this network of problematisations that is captured by the inclusion map which may be thought of as a mechanism of mutual cognitive adjustment. When a domain is looked at in this way it becomes obvious that there can be no single preferred way of describing its contents. Other chains of translation will branch out (indefinitely) if one starts from a different point in the network. Some of these chains are made visible by inclusion maps.

The multi-polar structure of the map of the domain of dietary fibre also reveals 'subterranean' connections. 'Voie biliaire' (biliary tract) is a peripheral issue, but it connects the central pole of 'alimentation' with the mediator 'tumeur' which is included in a different pole ('colon'). Two different sets of simplifications are connected by a passageway that is, at least for the time being, a minor problematisation. The latter might grow, however, and rise on the map thereby leading to fusion between the two general themes which at present have separate hierarchies of simplifications.

The inclusion map for dietary fibre for the later period of 1977–8 (see Figure 7.9 which depicts the top part of this map) reveals that fusion has indeed taken place, though this did not result from an increase in the importance of the passageway through 'voie biliaire'.

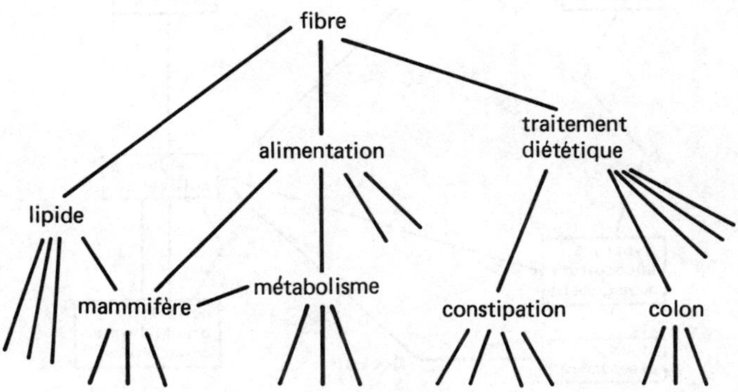

FIGURE 7.9 Inclusion map for dietary fibre research, 1977–8

A single pole, 'dietary fibre', now describes the domain and its subdomains. 'Alimentation' and 'traitement diététique' are much more firmly linked to diet and the role of fibrous materials in metabolism and disease than was the case in the earlier map.[7] The arrangement of simplifications in the domain has changed, and the head of the scientific agency will now get different answers from his scientific director if he enquires about its state.

The proximity maps of the dietary fibre area turn out to be rather uninformative (see Callon, Courtial, Turner and Bauin, 1983). They reveal a few clusters of peripheral key-words that also appear at the bottom of the inclusion map. In fact, in a strongly hierarchised domain the proximity map is often rather uninteresting. By contrast, in domains with an 'exploded' network of problematisations (see Chapter 11) – those with many peripheral words subsumed directly under a central pole with no intervening structure of mediators – it is often the inclusion maps that are uninformative, while the proximity maps may be richly patterned and reveal a number of densely connected clusters linked to each other, often by more tenuous connections. Thus Figure 7.10 shows the main outlines of a proximity map constructed from the database 'aquaculture' as provided by the Food and Agriculture Organisation (FAO).[8] This file contains about 1600 documents published in 1979, the period which we will consider here. The map breaks down into six interconnected clusters of words: 'fishery and aquaculture management', 'breeding and cultivation', 'pathology/diseases', 'feeding and nutrition', 'stock assessment' and 'pond and brackishwater aquacul-

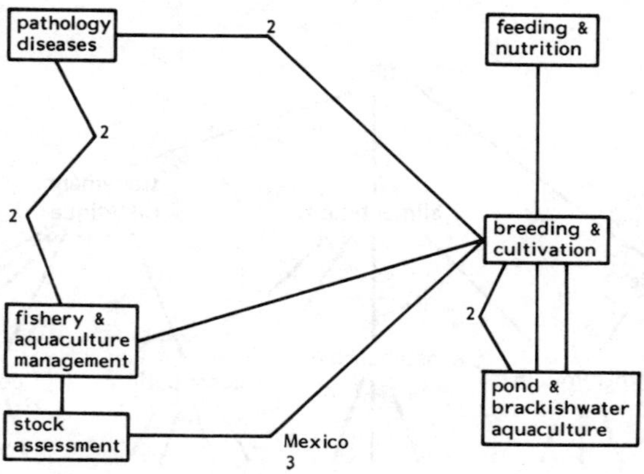

FIGURE 7.10 Proximity map of aquaculture, 1979

ture'. (The labels used here are not necessarily key-words, but are our way of summarising the common features of the problematisations within a cluster.)

One striking aspect of the clusters is their heterogeneity. Consider the cluster 'breeding and cultivation' (Figure 7.11). Here there are culture techniques (such as 'offbottom culture'), effects that may be related to theoretical interpretations (for instance, 'temperature effects'), natural entities (e.g. '*Crassostrea virginica*'), artefacts ('culture tanks') and countries ('Brazil'). There is no easy and obvious way to classify this

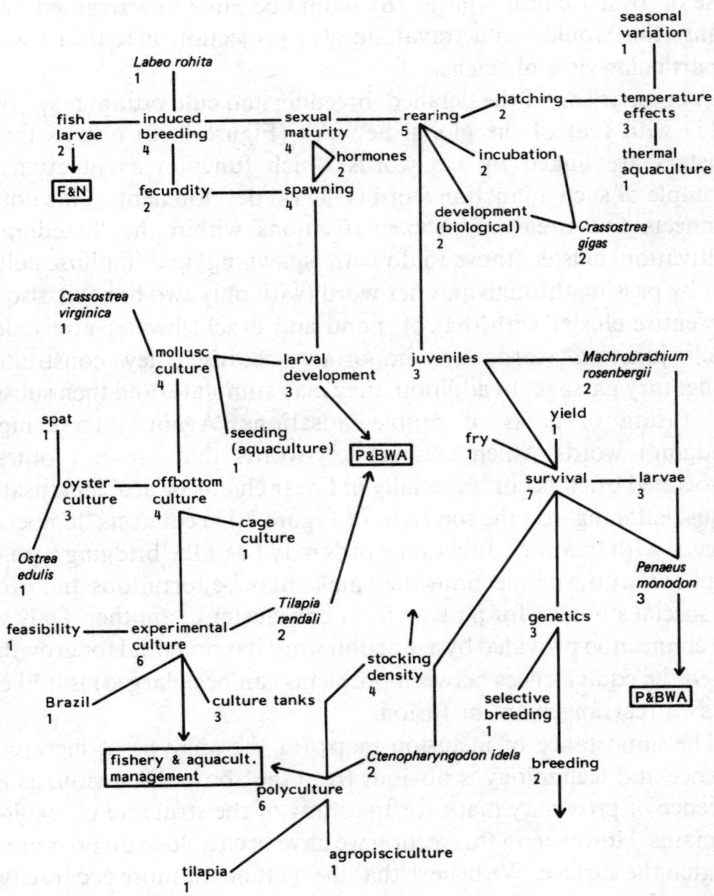

FIGURE 7.11 Proximity map of breeding and cultivation cluster

heterogeneity, but that is as it should be. Networks of problematisations cannot be captured by the clean conceptual distinctions preferred by many theorists of science. Science is rather made up of bits and pieces, and the fact that these proximity maps reflect its heterogeneity is an advantage rather than a problem. For instance, 'Brazil' is involved in the study of 'experimental culture' and the latter is a passageway between 'culture tanks' and 'feasibility'. Phrased like this, the heterogeneity is obvious. However, it is important to avoid the trap of reconstructing such linkages in a purely intellectual way. Thus a claim that the link between two research areas, 'experimental culture' and 'offbottom culture', exists because these two problem areas are being studied for the case of 'oyster culture' in 'Brazil' would be quite unwarranted. Such a translation would be a betrayal, simply a projection on to the network of a particular view of science.

A comparison of the detailed 'breeding and cultivation' map (Figure 7.11) with that of the global network (Figure 7.10) reveals that the clusters are linked by key-words which function as junctions. An example of such a junction word is 'larval development'. This not only connects two areas of problematisations within the 'breeding and cultivation' cluster (those to do with 'spawning' and 'mollusc culture') but by passing through another word (with only two links) it also links the entire cluster with that of 'pond and brackishwater aquaculture'. Such junction words are important because they constitute an obligatory passage. In addition, they may stimulate (and then subsume) the fusion of areas of problematisations. Again, intervening (or bridging) words which connect key-words that are not otherwise associated often occur, especially in linear chains of problematisations. Thus 'hatching' (on the top right of Figure 7.11) connects 'temperature effects' with 'rearing'. Junction words may be of the bridging type. This implies that the connections they make may be fortuitous and provide no special stimulus for passing from one cluster to another. Only when the connection provided by a junction word has potential for growth (i.e. when the equivalences between problems can be enlarged) is it likely to lead to rearrangement or fusion.

The importance of inclusion maps for the analysis of hierarchy in science and technology is obvious from the above discussion, as is the salience of proximity maps for the study of the structure of 'exploded' domains. However, in this section we have been able to do no more than scratch the surface. We believe that the method, or more accurately, the range of methods of co-word analysis have considerable potential for the analysis of the dynamics of problematisation in science. In its pursuit of the qualitative by other means, the approach makes possible an analysis of the emergent effects of the interaction between a population of scientific actor-networks. This potential is further explored for a number of case studies in the chapters that follow.

NOTES

1. Co-citation is far from being the only method used to measure science. For an indication of the range of possibilities see, e.g., Elkana *et al.* (1978), Ellis, Hepburn and Oppenheim (1978), Freeman (1983), Garfield (1979), Gilbert and Woolgar (1974), Machlup (1972), Menard (1971), Narin *et al.* (1976), National Science Board, *Science Indicators* (1973, 1975, 1977, 1979, 1981), OECD (1982), Pavitt (1978, 1982), Schmookler (1966), Small (1977), Price (1963) and the journal *Scientometrics*.
2. In Chapter 5 we discussed another occurrence of networks of problematisation: those to be found in individual papers.
3. The kinds of words chosen for inclusion depend upon the users envisaged. These include scientists working in related areas, other scientists, documentalists, students, industrialists and sometimes also journalists and civil servants.
4. For further details see Chapter 11.
5. On the notion of black boxes see Callon (1981b), Law (1984b).
6. Figure 7.8 is reproduced with permission from *Social Science Information*, **22** (1983).
7. For details of the maps see Callon, Courtial, Turner and Bauin (1983).
8. This is described in more detail in Chapter 8.

8 Aquaculture: A Field by Bureaucratic Fiat

SERGE BAUIN

INTRODUCTION

In France, the CNEXO (the French council that is concerned with all ocean-related research and development, except fisheries) and other government agencies have made a concerted attempt to create a field of aquaculture. GIDA, (Groupe Interministériel pour le Développement de l'Aquaculture – Interministerial Committee for the Development of Aquaculture) was established to coordinate the activities of a range of interested ministerial services and to centralise information. In addition, Regional Delegates to Aquaculture (Délégués Régionaux à l'Aquaculture) were created. Their double role is to collect information in their own region and to communicate both this information and decisions made by national agencies to the different parties concerned. Thus, results from autonomous localities are brought together and made available to all those involved. Finally, laboratories (like the Institute National de la Recherche Agronomique (INRA) laboratory of fish physiology) were moved from Paris suburbs to more 'maritime' locations (for instance in Brittany), and were required to consider more applied questions (like antibiotic treatment and vaccination at trout and salmon farms in the estuaries of Brittany) in addition to their more usual basic research on pathogenic agents and fish genetics (Anon, 1981).

By contrast, researchers have often been sceptical. Definitions of what the field 'aquaculture' should consist of invariably lead to endless discussion. In a sense, then, researchers maintained their own, often local approaches, even if they were prepared to accept a general label such as aquaculture. In sum, it is often said that aquaculture is not a real field but rather a bureaucratic category.

Co-word maps should be able to trace such a situation in detail and identify changes over time. To this end, a file was drawn from the DOCOCEAN database of the FAO. This database is worldwide but we assumed that in other countries there is no more (and probably less) integration than exists in France. DOCOCEAN is indexed through key-words extracted from a controlled English language lexicon.

The boundaries of this file are consistent with the principles presented in Chapter 11. For this analysis, articles about aquatic culture in general were extracted from this database. In this way all articles containing the word 'culture' were retained whether this word appeared alone or not. Thus the word 'aquaculture' does not constitute the only central pole in our maps: articles containing words like 'crustacean culture' or 'mollusc culture' are only rarely indexed by 'aquaculture'.

Only the years 1979 and 1981 were used for this presentation. These two files amply reflect our main results. They contain approximately the same number of articles (1574 and 1736 respectively) thus facilitating direct comparison between the two periods.

INCLUSION MAPS

To conserve a certain degree of legibility, each inclusion map was drawn on to about fifteen pages for each period. Because of space restrictions they will only be partially presented and interpreted here: the first page (among 14) of the 1979 inclusion map is the only one represented in full (Figure 8.1); Figures 8.2 and 8.3 abstract the main observations made in both years (they list the central poles and the main mediator words included), and details from these maps are provided in support of the main points made in the text (Figures 8.4 to 8.7).

Unlike the results obtained from other files (e.g. the dietary fibre file used in Chapters 7 and 11), the inclusion maps of aquaculture show that there is very significant dispersion in this field. This can be indicated by the number of 'cycles' and mediator words present in any chosen graph or portion of a graph: the more the number of cycles and mediator words the less the degree of dispersion.[1] On the first page of the 1979 inclusion map (see Figure 8.1) we find the central pole 'fish culture' and three relatively important mediator words: 'aquaculture development', 'fresh-water aquaculture' (which includes 'fishery development' and 'inland fisheries'), and 'growth'. Although these six 'structuring' words are distributed over approximately forty peripheral issues, there are only ten cycles. The other pages of this map (not shown here) are even less structured.

Let us examine 1979 and 1981 in greater detail. The decomposition of

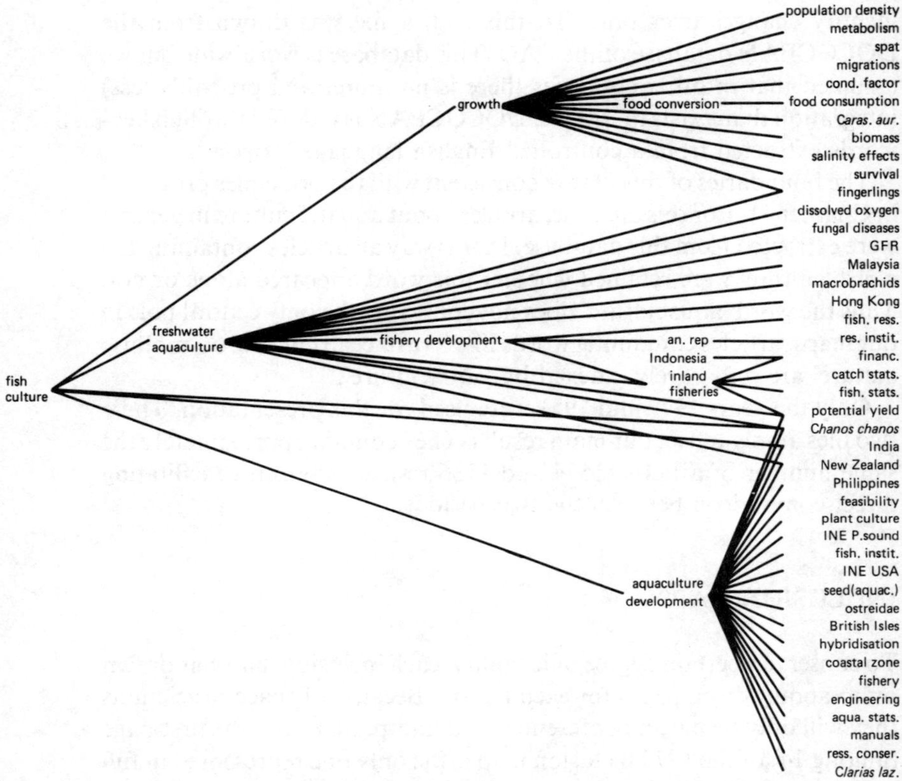

FIGURE 8.1 Inclusion map (detail), 1979

key-words into central poles and main mediator words is shown in Figure 8.2 for 1979 and Figure 8.3 for 1981.

For the first period (1979), the first central pole ('fish culture') is broken down into ten main mediator words (see Figure 8.2). It is surprising to find that a certain species '*Salmo gairdneri*', or the rainbow trout, which has been massively bred in freshwater since the nineteenth century, and less extensively in Norway's seas since the 1950s, is a high-frequency mediator word. Despite the frequency of research on this species, it does not create a very interesting structure (see Figure 8.4 which is a detail from the 1979 inclusion map which concerns this mediator word). Although seven words are included in the pole '*Salmo gairdneri*', there are no cycles and only two passageways toward other poles. A qualitative study has revealed that, like its cousin the salmon, this species has attracted the interest of industrialised countries because

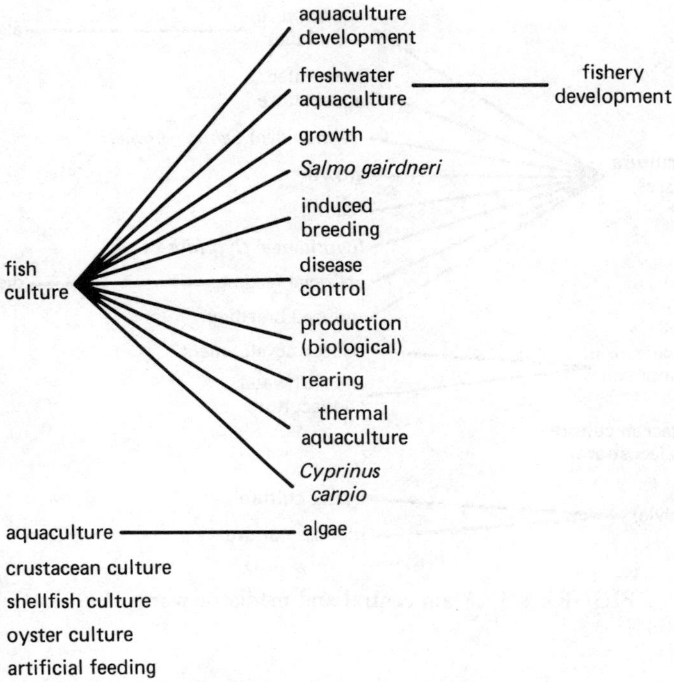

FIGURE 8.2 Main central and mediator words, 1979

of its commercial value. However, as the presence of the key-word 'cost analysis' confirms, the prospect of breeding this species in a profitable manner seems to face various problems (Anon, 1982). In addition, as is the case for other salmonids, critics of an 'aquaculture for the rich' have often denounced the wasteful protein reconversion involved in breeding this species. The frequency of research on trout reflects a certain degree of opposition between research destined for typically Western applications and classical writing on aquaculture which has a somewhat 'Third World' perspective (e.g. Haberkorn, 1975; Anon, 1978; Lelarge, 1980).

The second central pole for 1979 in Figure 8.2 is 'aquaculture'. Many of the words which this brings together are quite general. They can be qualified as having a 'delocalising' effect: their presence reflects the intention, already mentioned in the introduction,of bringing together local results. Words such as 'conference', 'education', and 'historical account' serve this purpose. The most structuring mediator word included under 'aquaculture' is 'algae' (see Figure 8.5 showing the

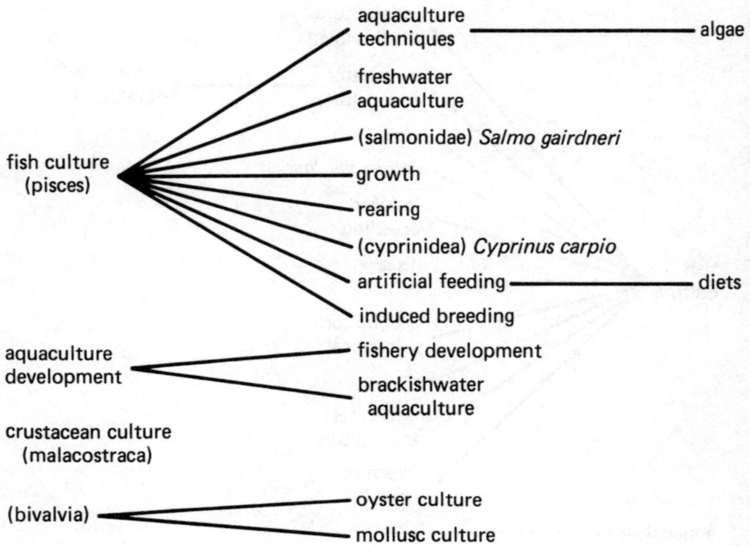

FIGURE 8.3 Main central and mediator words, 1981

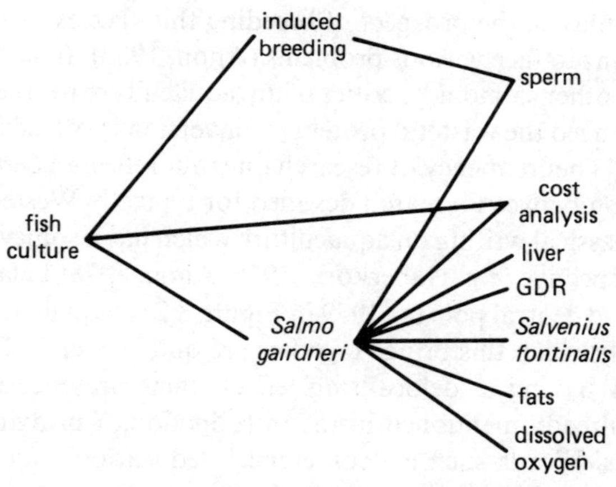

FIGURE 8.4 Detail from 1979 inclusion map (*Salmo gairdneri*)

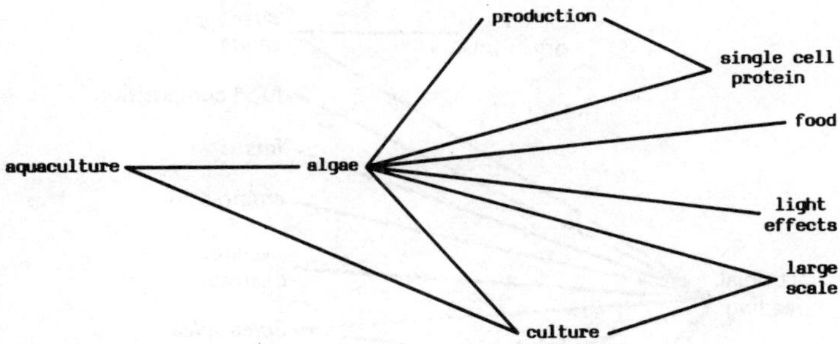

FIGURE 8.5 Detail from 1979 inclusion map (algae)

relevant detail from the 1979 inclusion map): it brings together six words for which there are two cycles and a second direct link with 'aquaculture'. These were rather marginal research projects and the recent development of programmes of aquaculture has no doubt enabled these projects to be more effectively translated. Although the algae researchers may, in the past, have specialised in what used to be insignificant and uninteresting vegetable species, they are now considered to play a role in the development of the field of aquaculture.

The two sub-fields corresponding to the central poles 'crustacean culture' and 'oyster culture' (or 'shellfish culture') are quite independent from research on fish or algae (that is, they have no link with the rest of the map). Although, like the rest of the field, these two sub-fields are poorly structured and include the names of many species, they differ greatly from the other central poles because research on crustaceans or molluscs is often linked to specific geographical locations where the experiments are performed. They are, literally, localised. This is especially the case for research done in Japan, Thailand, and various North American states. Further, the breeding of crustaceans (lobster, shrimps, etc.) is a relatively new idea. The presence of the key-word 'laboratory culture' confirms this observation. Thus the breeding of these species is mostly experimental and results prudently refer only to the locations where these results were obtained.

There is a final area of research which is almost completely autonomous within the inclusion map (see Figure 8.6, which depicts relevant details from the 1979 inclusion map). This group is simplified by the central pole 'artificial feeding'. It consists of research that is principally concerned with feedstuffs and uses key-words like 'feed composition', 'nutritional value', 'chemical composition', etc. The

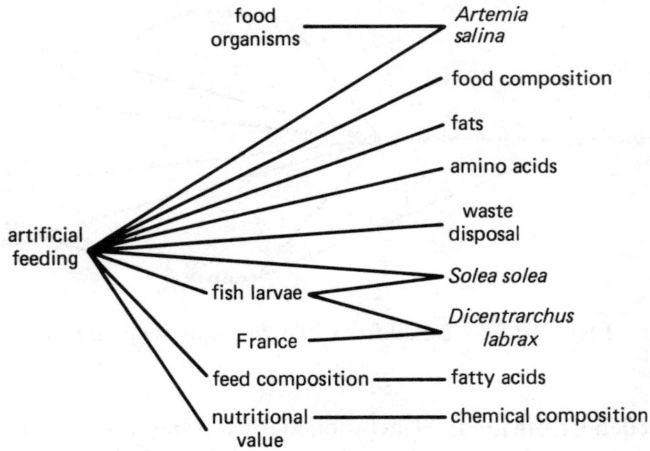

FIGURE 8.6 Detail from 1979 inclusion map (artificial feeding)

structure of this group is somewhat less dispersed than it is for the field as a whole. Although there is only one cycle, there are three mediators.

With the exception of 'algae' and to a lesser degree 'artificial feeding', the 1979 portrait of aquaculture reveals an exceptional dispersion of the research on marine culture in general. Even sub-fields like 'crustaceans' and 'molluscs' are poorly structured. We shall now see whether it is possible to detect any significant change in the field as a whole.

The same dispersion of the structure is apparent in the inclusion map of 1981. This turns out to be a characteristic of all the periods that were studied.

First, as has been pointed out by the indexers of the DOCOCEAN database, the addition of the word 'pisces' to almost all the documents concerning fish turns out to be a simple artefact of indexation. It is thus an unimportant development which warrants no further analysis.

The breakdown of the field into central poles and main mediator words remains relatively stable in 1981 (Figure 8.3). For example, words like 'growth', 'freshwater aquaculture', or 'induced breeding' remain under the central 'fish culture' pole. In addition, research on fish, molluscs and crustaceans retains the same degree of mutual autonomy.

None the less, changes of differing degrees of significance have occurred. One example of a minor transformation is that the mediator word '*Salmo gairdneri*' has been replaced by 'salmonidae'. Other salmonid species are slightly better represented although the rainbow trout still occupies an important position. Among the more significant changes, it should be noted that the central pole 'aquaculture' has disappeared. It has been replaced by two new poles: (a) 'aquaculture development' which no longer concerns fish alone; and (b) 'aquaculture techniques' which now includes the breeding of algae. The pole 'aquaculture development' covers all the 'general' key-words from the 1979 inclusion map. For example, 'conference', 'sociological aspects', or 'historical accounts' are found under this pole in 1981. Further, through the mediator 'fishery development', it enables researchers who work upon different themes related to fisheries to find a place within the system of marine culture research. It is interesting to note that this sub-field is more structured than the rest of the file. It has several transversal lines around 'annual reports' and 'fishery institutions'.[2] The fact that the sub-field has existed longer and is related to a mass production activity might explain the greater degree of structure.

The word 'artificial feeding' loses its status as a central pole (see Figures 8.2 and 8.3, and Figure 8.7 which is a relevant section from the 1981 inclusion map). It is included under 'fish culture' in the 1981 map. In itself, the appearance of this link is not very significant because the inclusion coefficient probably passed from a value slightly lower than the threshold to one that was higher. However the activities which are related to this sector have undergone a notable and visible change. The maps reveal an evolution from the field of feedstuffs research, which is undertaken by industrial laboratories, to research using a perspective closer to dietetics. Thus the two important mediators 'diets' and 'nutritional requirement' as well as the words 'metabolism' and 'animal nutrition' appear on the 1981 map. Of course this does not mean that a total revolution has taken place. Neither feedstuffs research (attested by the words connected directly to 'artificial feeding' on the top right of Figure 8.7), nor the animal feed industry have suddenly disappeared. Nevertheless, the reorientation is clearly apparent in the maps.

Research on crustaceans, molluscs and bivalves has undergone few modifications between 1979 and 1981. Like the 'pisces' pole, the poles 'malacostraca' and 'bivalvia' are only artefacts of indexation. The appearance of these poles is due to an improvement in the vocabulary used by the indexers of the FAO database. Here again these two research fields are exceptionally dispersed: almost no transversal links structure the maps.

In this section we have drawn a general portrait of the field of aquaculture. Using the proximity maps, it is possible to uncover the specific issues which are studied by researchers as well as the manner in which these are joined together under the central poles revealed by the inclusion maps.

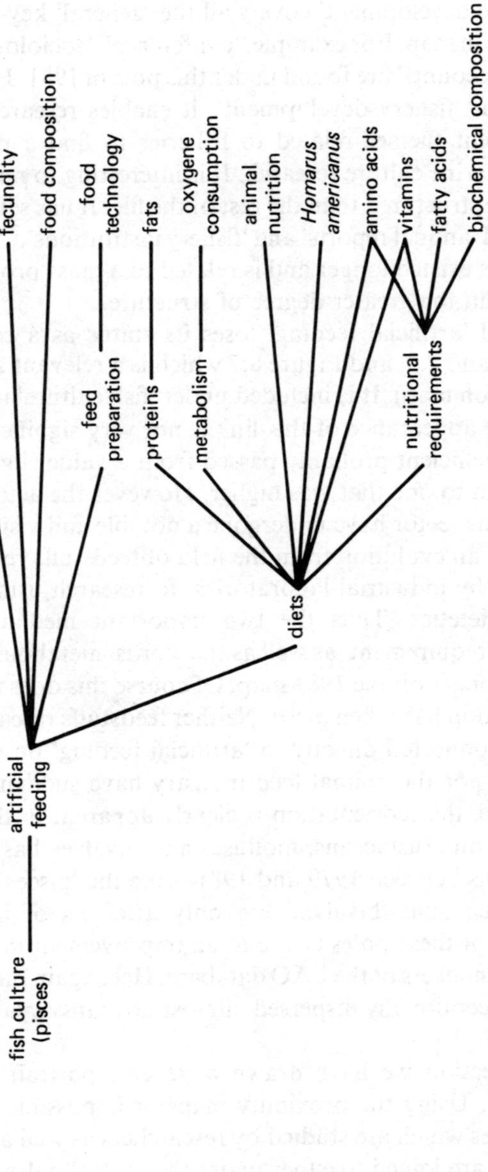

FIGURE 8.7 Detail from 1981 inclusion map (artificial feeding)

PROXIMITY MAPS

Both 1979 and 1981 proximity maps have been drawn (Figures 8.8 and 8.9). In order, again, to conserve a certain degree of legibility, clusters of highly related words have been abstracted in titled squares such as 'feeding and nutrition' or 'breeding and cultivation'. Chains linking these clusters are drawn. They are composed of words which are spelled out only if they are related to three or more (five or more for 1981) other words. The number of links is indicated for each of these words. Remaining words belonging to these chains are only marked by this number. Finally, words linked to the rest of the network by a single link have been omitted.

The clusters which are discussed in this section are presented in detail in Figures 8.10 to 8.15 where no words are omitted.

Let us finally recall that in the proximity maps, unlike the inclusion maps, the links are symmetrical; thus the relative 'geographical' position of the words is of no importance: the only significant feature of the graph is the presence or the absence of a relation between two words.

Let us examine a few of the issues which were identified in the 1979 map and indicate the way in which these have evolved in the 1981 map.

There is a cluster referring to research on feeding and nutrition (see Figure 8.10). As we have noted, these issues are simplified under the pole 'artificial feeding' on the 1979 inclusion map. This same cluster is considerably extended in 1981 (Figure 8.11). It is also more structured: with an unchanged threshold the average number of links per word

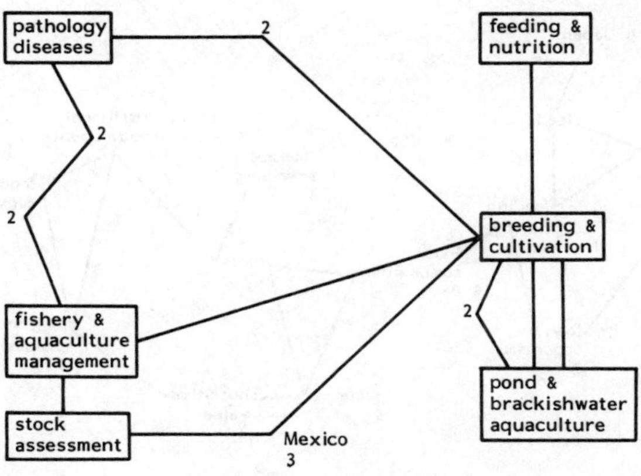

FIGURE 8.8 Proximity map, 1979

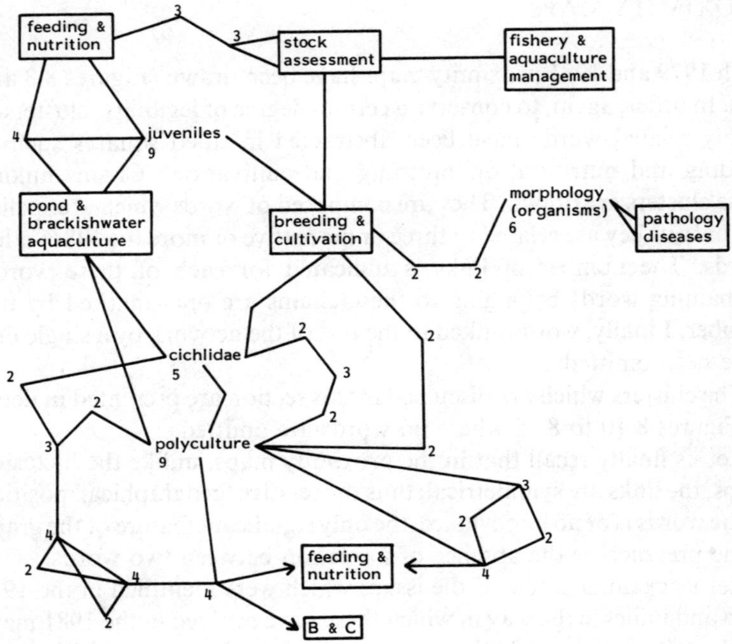

FIGURE 8.9 Proximity map, 1981

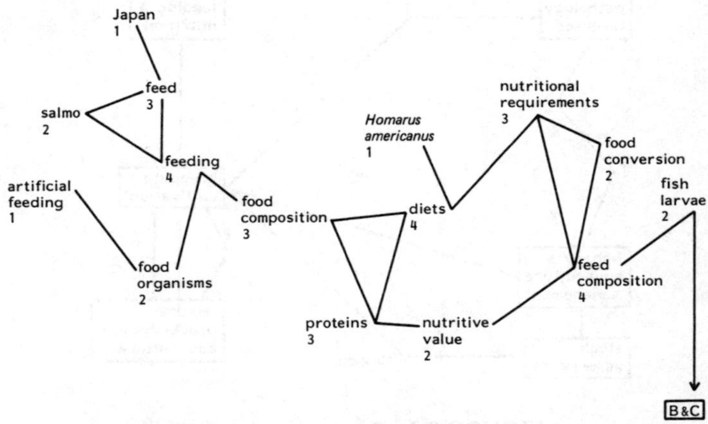

FIGURE 8.10 Feeding and nutrition cluster, 1979

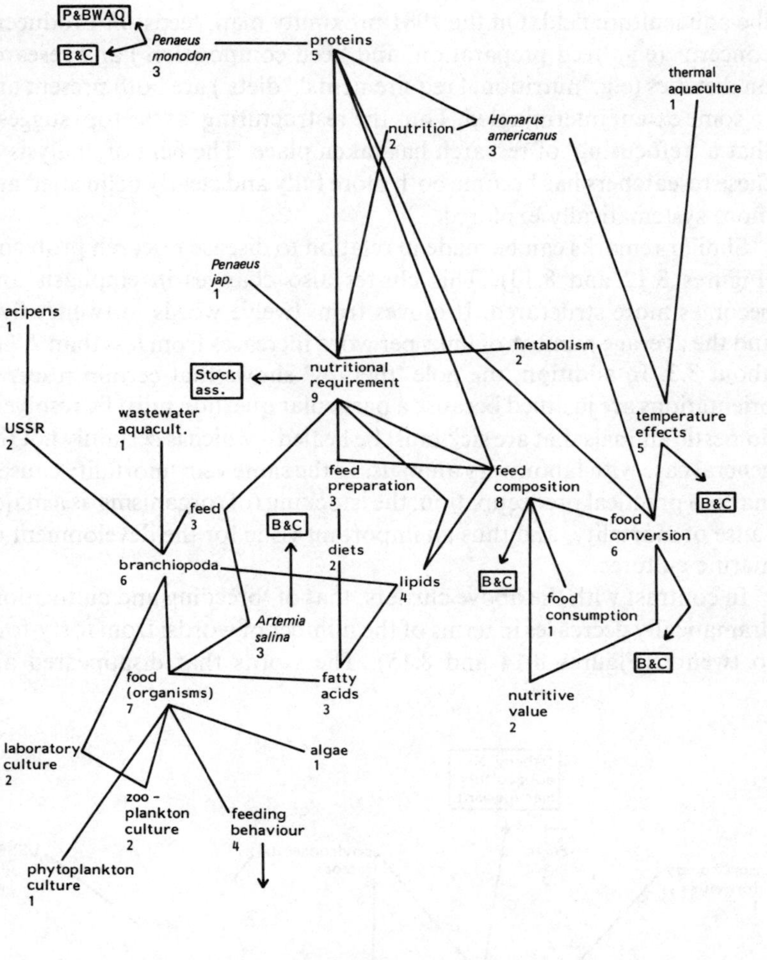

FIGURE 8.11 Feeding and nutrition cluster, 1981

increases from 2.5 to 3.2. Two features of the inclusion maps must be recalled in relation to this evolution. First, the pole 'artificial feeding' which is a central pole in the 1979 inclusion map is included under 'fish culture' in 1981. In addition, the words which this pole brings together reveal an evolution from research on feed itself to research using a perspective closer to that of dietetics. The evolution on the proximity maps is quite different. On the 1979 proximity map, the 'dietetics' orientation is already visible (through e.g. 'diets', 'nutritional requirement'). Thus dietetics did exist, but was still of minority interest in the aquaculture field. On the 1981 proximity map, feedstuff producers' concerns (e.g. 'feed preparation' and 'feed composition') and research on dietetics (e.g. 'nutritional requirements', 'diets') are both present and to some extent intermingled. Thus the restructuring 'at the top' suggests that a 'refocusing' of research has taken place. The field of analysis of these researchers has become both more fully and clearly delineated and more systematically explored.

Similar remarks can be made in relation to disease research problems (Figures 8.12 and 8.13). This cluster also changes in emphasis and becomes more structured. It moves from twelve words to twenty-five, and the average number of links per word increases from less than 2.7 to about 3.3. In addition, the pole 'therapy' shows that certain research orientations are justified because a particular question must be resolved: domestic animals that are sick must be healed – which is certainly not the general case with laboratory animals. In the same vein 'mortality causes' marks a practical preoccupation: the 'stocking (of) organisms' is a major cause of mortality, and thus an important issue for the development of marine cultures.

In contrast with the above clusters, that of 'breeding and cultivation' dramatically decreases in terms of the number of words: from forty-four to twenty (Figures 8.14 and 8.15). The words that disappeared are

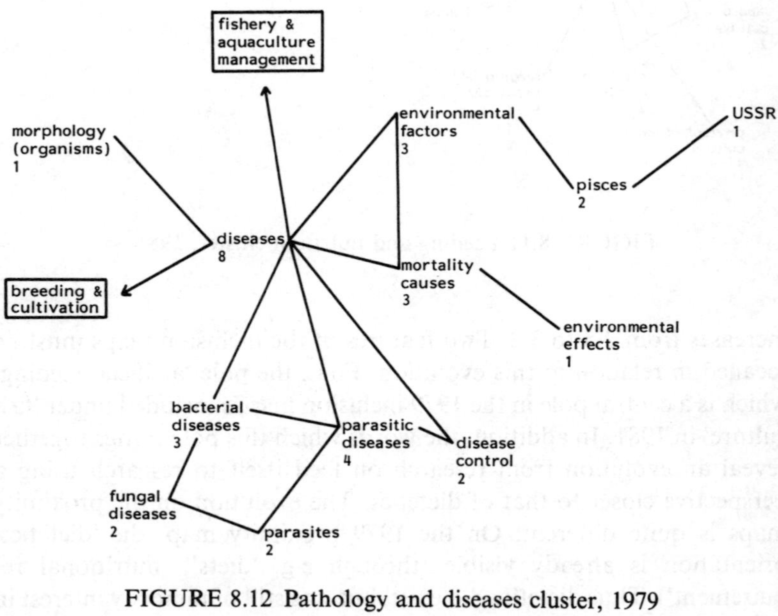

FIGURE 8.12 Pathology and diseases cluster, 1979

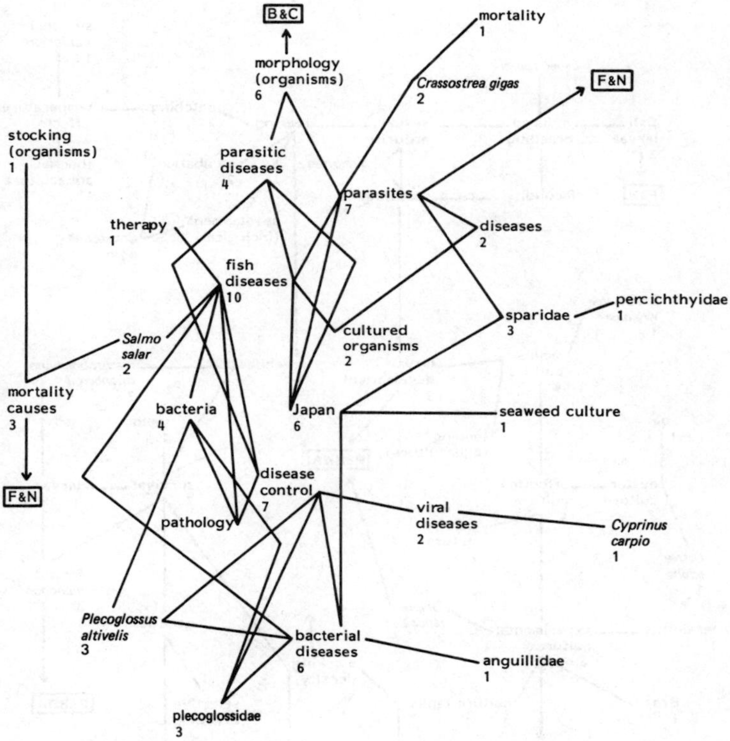

FIGURE 8.13 Pathology and diseases cluster, 1981

mainly the names of certain types of culture ('experimental culture', 'offbottom culture' etc.) and associated words. Some of them remain as poles (and even sometimes as central poles) on the inclusion maps. Others, like 'polyculture' are at the crossroad of the chains relating the clusters (see Figure 8.9). Thus these words begin to be delocalised. One could say that they no longer concern certain specific researchers or experiments, but become the potential applications of all the research domains as identified by the clusters. On the other hand, the average number of links per word of the 'breeding and cultivation' cluster increases from less than 2.5 to 3.2, thus becoming more structured while smaller.

More generally, for the complete proximity maps, the average number of links per word has increased from 2.33 to 2.95. This might be the mark of the beginning of the integration of the whole field, although caution is required since there is only one indicator.

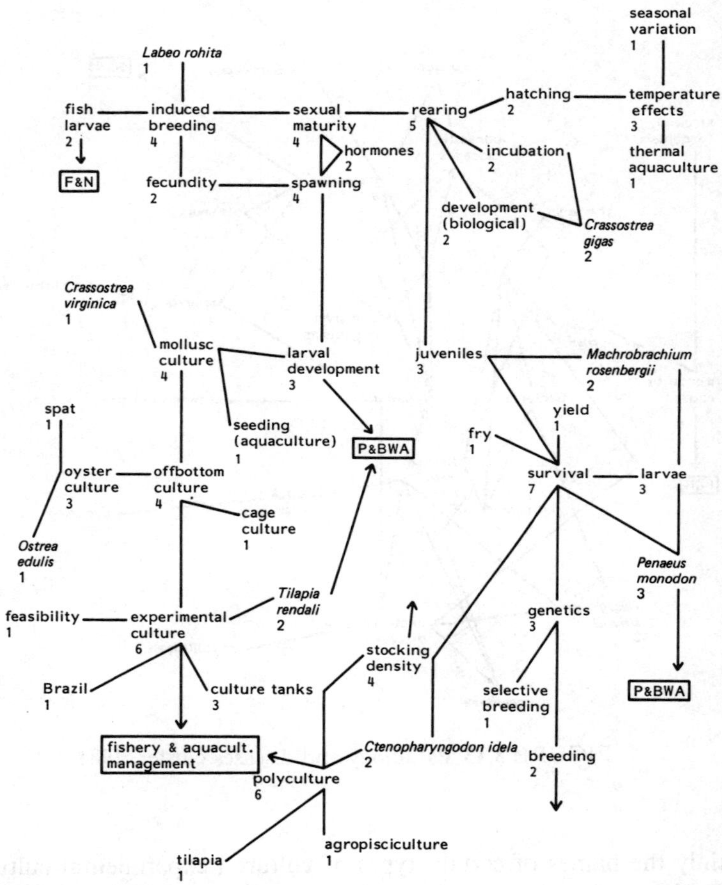

FIGURE 8.14 Breeding and cultivation cluster, 1979

CONCLUSIONS

The inclusion maps reveal that the field of aquaculture is remarkably dispersed. There are numerous links between very specific, infrequent words and the central poles; the only meaningful structures detected are a rough separation into subspecies (crustaceans, molluscs and fish), very general problems ('aquaculture development'), and a few specific research problems ('growth' or 'rearing'). By contrast, the proximity maps reveal a very high degree of structuring in which no one issue is

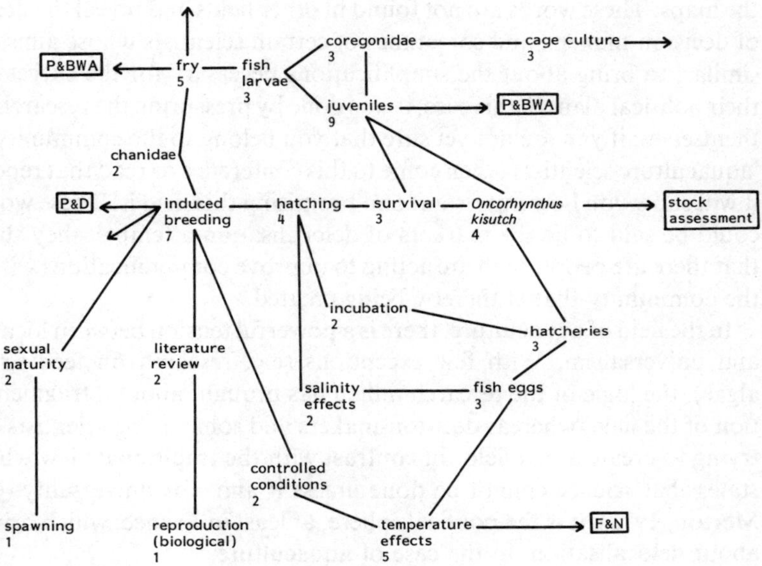

FIGURE 8.15 Breeding and cultivation cluster, 1981

totally disconnected from the rest of the field. Although these observations are partial it is none the less possible to draw a number of conclusions.

First, as we have noted, aquaculture, as a unified field of preoccupations and objectives, does not exist outside the political influence of certain decision-makers. Thus, and this can be seen in the maps, researchers have maintained their respective approaches even though they have agreed to link their work to a few of the central poles which have been identified in the maps. These researchers have remained local and locally connected (although they may have been, at the same time, in a different position in fields other than aquaculture). Even if the elaborated knowledge of these different scientists is placed in black boxes, these boxes are hardly related to the outside world. There is no hierarchisation which would enable other scientists or companies to use these results when they undertake other research projects or develop an industrial product. In the same vein, as earlier noted, the names of geographical locations occurred frequently in this file, many results thus remaining literally 'localised'. This characteristic has not been found in any other Leximappe study. In addition, words like 'conference', 'annual reports', 'historical accounts' or 'sociological aspects' appear on the maps. These words are not found in other fields and reveal the desire of decision-makers, and sometimes of certain scientists whose aims are similar, to bring about the simplifications necessary for the success of their political plan. In this case, it was done by pressuring the researchers themselves: if you are not yet sure that you belong to the community of 'aquaculture scientists', then come to this conference or read that report; it will show you how important and promising this field is! These words could be said to be the markers of delocalisation attempts: they show that there are people who are acting to improve communications within the community that is thereby being created.

In the field of aquaculture, there is a powerful tension between locality and universalism. With few exceptions (e.g. research on feeding or algae), the logic of the research milieu has brought about a fragmentation of the field, whereas decision-makers and some of the scientists are trying to create a real field. In contrast with the traditional view which states that science cannot be done unless it aims for universality (e.g. Merton, 1973), it is the political sphere, at least in France, which brings about delocalisation, in the case of aquaculture.

However, certain changes in structure that have been noted reveal an appreciable degree of evolution. Perhaps policy has some influence or perhaps the scientific tendency toward universalism is at work. The changes in disease research, together with the move of the INRA lab illustrate the first. The fact that research on feeding is less exclusively a practical concern of feed producers, but has also been related to general research into diets can be seen as an example of the tendency toward universalism.

In the proximity maps further changes can be identified. The changes in the 'feed' area can now be traced in detail. Some areas are constant, others lose many words (e.g. breeding and cultivation), still others shift emphasis (pathology and diseases).

Thus co-word mapping is a powerful tool that makes it possible to trace the structure and the evolution of a socio-cognitive network. This is equally true for those cases where there is no global or paradigmatic structure. Thus, where the co-citation method may be unable to describe the aggregation of 'specialities' into a coherent field, co-word analysis is able to reveal the dispersion of a field by comparing poorly structured inclusion maps with highly structured proximity maps.

NOTES

1. In graph theory, the notion of cycle is in opposition to that of trees. Thus in Figure 8.1 the fragment composed of 'fish culture', 'aquaculture development' and '*Chanos chanos*' represents a cycle. In fact there are two ascending paths which join '*Chanos chanos*' to 'fish culture'. To obtain a tree one only has to remove the line linking '*Chanos chanos*' to 'fish culture' or 'aquaculture development'. For more details on graph theory, see Gondran and Minoux (1979) and in relation with the Leximappe technique see Chapter 11.
2. Note here the intrinsic relativity of the hierarchies obtained by the Leximappe technique. If the file had been formed around problems concerning fisheries research, then aquaculture would have been included as a sub-part in the larger network of problematisation.

9 State Intervention in Academic and Industrial Research: The Case of Macromolecular Chemistry in France

WILLIAM TURNER and MICHEL CALLON

The increasing use of science policy programmes to promote and orient research activities underlines the need for analytical tools which can contribute to their elaboration and evaluation. There is a wide variety of science policy objectives and a corresponding variety of criteria that can be used to measure the impact of initiatives for further research. Two classes of policy initiative are, however, of particular importance. The first aims to promote, develop and improve existing activities without necessarily changing their content. The second seeks explicitly to modify the content of a field by introducing new themes of study, creating links between previously unrelated sectors of activity, or by trying to promote the emergence of new problem areas. In the first case, decision-makers, whether they be hybrid or homogeneous (Weingart, 1982), generally need only those output indicators that allow them to appreciate the quality of a laboratory's (Martin and Irvine, 1983) or a firm's production, for a given set of inputs (OECD, 1982). In the second case, decision-makers are confronted by a series of strategic questions: What are the research problems at present being studied? What themes of common interest can be defined which will lead to closer ties between laboratories working in different areas? How can one be certain that new and important fields are not being completely ignored by the national research community? In order to contribute to the discussion which this type of question raises, analytical tools which make it possible to consider the content of scientific and technical research are needed (see Chapter 7). Output measures are no more useful in this case than are macro-economic indicators when the goal is to define specific policies for sectors of industry. This chapter is a follow-up to a study which was published in 1979 about a Concerted Action Programme which was designed to improve France's position in the field of macromolecular chemistry (Callon, Courtial, Turner, 1979). The programme was established in 1963 by the Délégué Général à la Recherche Scientifique et Technique (DGRST). Its goal was, in particular, to coordinate university and industrial research in the sector.

In that earlier study, we analysed the scientific content of the

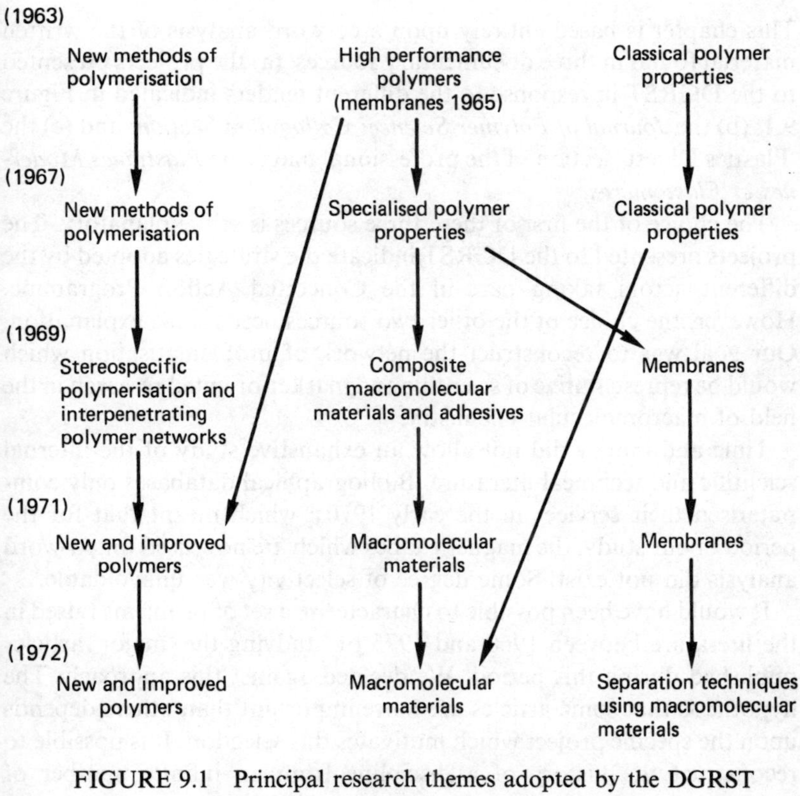

FIGURE 9.1 Principal research themes adopted by the DGRST

programme, and changes in this content between 1963 and 1975. Figure 9.1 indicates the principal research themes adopted by the DGRST in each of the five tenders that it sent out over this period. The arrows drawn between these themes indicate that, despite changes in the names of the categories used to build bridges between university and industrial laboratories, the subject content of the programme remained relatively stable (Bouzat and Grislain, 1979).

In this chapter we are interested in the cumulative impact of the Concerted Action Programme. We will use co-word analysis to study the way in which the science policy of the DGRST structured the research field of macromolecular chemistry in France.

DATA AND INDEXING PROCEDURES

Documentary Sources

This chapter is based entirely upon a co-word analysis of the written material found in three documentary sources: (a) the projects presented to the DGRST in response to the different tenders indicated in Figure 9.1; (b) the *Journal of Polymer Science: Colloquium Section*; and (c) the 'Plastics Digest' section of the professional magazine *Plastiques Modernes et Elastomères*.

The choice of the first of these three sources is self-explanatory. The projects presented to the DGRST indicate the strategies adopted by the different actors taking part in the Concerted Action Programme. However, the choice of the other two sources needs some explanation. Our goal was to reconstruct the network of problematisation which would be representative of scientific and market-oriented research in the field of macromolecular chemistry.

Time and money did not allow an exhaustive study of the internal scientific and technical literature. Bibliographical databases only computerised their services in the early 1970s, which meant that for the period of our study, the magnetic tapes which are now used for co-word analysis did not exist. Some degree of selectivity was unavoidable.

It would have been possible to characterise a set of problems raised in the literature between 1960 and 1975 by studying the 'major' articles published during this period. We decided against this approach. The hypothesis that some articles are more important than others depends upon the specific project which motivates this selection. It is possible to reconstruct the history of a discipline from an infinite number of viewpoints. Rather than select one of them, we opted for an exhaustive analysis of the full set of publications found in the colloquium section of the *Journal of Polymer Science*, on the one hand, and the industrial announcements concerning new products and plastic material applications found in the 'Plastics Digest' section of *Plastiques Modernes et Elastomères*, on the other hand.

The set of problems identified in the colloquium section of the *Journal of Polymer Science* presented an advantage from our point of view. The themes chosen for the organisation of scientific meetings are generally closely related to what is taking place on the research front. Researchers use these meetings to present their first results and confirm their interpretations before publishing (Meadows, 1974). Members of what Crane (1972) called 'invisible colleges' are present. According to Cole and Cole (1973) these people are the most influential in their discipline given the quantity and quality of their publications. Derek de Solla Price (1963) has underlined their mobility: they visit other laboratories and participate in scientific meetings in order to keep abreast of what is happening. Back in their own laboratories, they engage in research in order to remain 'competitive'. It is easy, then, to understand the importance of the set of problems identified in congress proceedings. These problems are often in an initial phase of development, and it is impossible to know whether the associations which are established between them will last or not. The same is true of the problems formulated in the projects submitted to the DGRST in response to a call for tenders. A parallel can thus be drawn between the action of international scientific meetings and the action of the DGRST in structuring scientific and technical activity. Both give priority to general themes of interest which are capable of combining a wide range of specific interests. These combinations result in networks of problematisation which encourage new orientations and new trends of research in the macromolecular chemistry field.

In the vocabulary of database producers, congress proceedings are classed as a part of the 'grey literature' (Keriguy, 1984). This type of literature is characterised by the fact that there is generally no well established circuit for its dissemination. The effort of the *Journal of Polymer Science* in the field of macromolecular chemistry is an exception which proves the rule: often the proceedings of a scientific meeting are not circulated outside the group of scientists who participated in the meeting.

The collection of information concerning innovations in industry is another example of the problems raised by this grey literature.

Professional magazines play an important role in overcoming some of the difficulties associated with the fact that the rules for the dissemination of information about innovations are much less well established in industry than in the scientific community. Some of the headings found in these magazines concern the fairs, exhibitions, state of the art and industrial announcements for a given field. We chose to analyse the industrial announcements concerning new products and material applications because, once again, these can be considered as a part of a general process leading to the recognition of the importance of specific products or applications over others (Allen, 1984). Each announcement contained a short but precise definition of a product, process or application which was being developed by a given firm. Also, this heading was present in each of the eleven annual issues of *Plastiques Modernes et Elastomères*, a fact which enabled us to follow the evolution of industrial preoccupations closely over the period studied.

Indexing Procedures

Of the 4360 documents in the data file, 41.3 per cent came from the *Journal of Polymer Science* (1843 articles), 39 per cent from *Plastiques Modernes et Elastomères* (1657 announcements), and 19.7 per cent from the DGRST (860 documents).

These different data sources utilised different vocabularies. The articles published in the *Journal of Polymer Science* made it possible to describe the scientific specificity of the problem raised. The industrial vocabulary was different: little information was given about the science behind an innovation, whereas the reasons for using it in certain market contexts were carefully described. For the DGRST, we were obliged to work with the summaries of the projects submitted for funding because the proposals themselves were not in the DGRST archives. In these summaries, general concepts were often used which served more to identify the various centres of interest of the groups taking part in the Concerted Action Programme than to give a detailed description of the scientific content or market implications of the proposal they presented.

Six students in the Masters Programme at the Strasbourg School for the Application of High Polymers indexed the data file. They were asked to respect the language found in each document by using, when possible, the concept employed by the authors themselves. They were also asked to avoid using general concepts which might have been suggested to them when reading a document, but which represented, in fact, an interpretation of its content. The goal of these indexing rules was to reduce the personal interpretation of the content of documents to a minimum. As a result of this effort, the specificity of the vocabulary found in each of the three data sources was preserved. The problem was then one of identifying the notions which served to build bridges between these different vocabularies, and thus between industry and research. This problem required an aggregation of data which was achieved by means of the hierarchical structure of the *Polymer Thesaurus* used by the Documentation Centre (CDST) of the French National Research Council (CNRS).[1] We replaced the specific key-words found in the vocabularies of our three documentary sources by more general terms which nevertheless indicated the various centres of interest of the documents analysed. In this way we were able to overcome the discontinuities of our data which resulted from the fragmented nature of our documentary sources. Had we been able to construct a more exhaustive data file containing the bridge-building literature between scientific meetings and industrial announcements, it would no doubt have been possible to avoid systematically substituting general aggregating terms for those designating specific centres of interest in the documents. In the same way, given that the DGRST programme was designed to establish links between science and industry, the fact that we were obliged to work with the summaries of the projects submitted to the DGRST, instead of the proposals themselves, is a factor explaining why we had to aggregate our data to the extent we did.

THE PROPOSALS SUBMITTED TO THE DGRST AND THEIR POSITION IN AN INDUSTRIAL/UNIVERSITY NETWORK OF PROBLEMATISATION

On an average, six key-words were used to index the summaries of the proposals submitted to the DGRST, eight for the industrial announcements in *Plastiques Modernes et Elastomères* and ten for the articles published in the colloquium section of the *Journal of Polymer Science*.

In order to study the cumulative effect of the DGRST's action in creating ties between industry and research, we first undertook a proximity coefficient analysis of the 860 summaries of the proposals addressed to the DGRST. A list of cluster/problem areas was identified which corresponded to groups of three or four associated words that were linked by a proximity coefficient greater than 10. A sentence was used to describe the subject matter of each word group. In the rest of the chapter we will designate these cluster/problem areas by the following numbers:

1 The production and assembly of thermoformed plastic sheets.
2 The study of polydienes in connection with coating and moulding techniques.
3 The study of the heat resistance of polyurethanes in connection with fireproofing techniques.
4 The study of complex catalysts in connection with the polymerisation of polyolefins.
5 The study of glueing techniques in connection with the use of threads to reinforce polyesters.
6 The study of how thermodynamic and surface properties affect material transformation.
7 Microstructural study of structure/property relationships with the aim of obtaining better rheological properties.
8 The study of membrane filtering techniques as a method for physical separation.

Our results indicate that university, public research and industrial laboratories bidding for money from the DGRST spontaneously organised their applications around a group of industrial processes. Material transformations, thermoforming, coating, moulding, fireproofing, glueing and filtering techniques are at the centre of the problems raised.

The next step was to see how the DGRST problem areas fitted into a network of problematisation generated by a proximity coefficient analysis of the documents found in the *Journal of Polymer Science* and *Plastiques Modernes et Elastomères*. This general data file contained 3067 documents.

We first used a high threshold (30) in order to identify the core definitions of problems found in the general file. The proximity graph obtained displayed a large number of small but totally connected components. Amongst them, two problem areas – coating and moulding techniques (DGRST cluster 2) and polymerisation techniques (DGRST cluster 4) – appeared as consolidated word groups similar to those which were found on the proximity maps for the DGRST sub-file.

We next lowered the threshold to 15. New clusters and words appeared on the graphs and, at the same time, the network of relationships between words became increasingly complex. Amongst the new problem areas identified, three of them corresponded to clusters, 1, 5 and 8 of the DGRST sub-file (thermoforming plastic sheets, reinforced polyesters and filtering techniques). The other DGRST problem areas (clusters 3, 6, 7) were located and simplified[2] in various connected components by single words which served to identify their point of entry into the global network. For example, cluster 3 concerning the heat resistance of polyurethanes was positioned in the general network at the 15 threshold by the word 'polyurethane' which was connected with 'foam materials' and 'lightweight plastics'.

The final step in our analysis was to lower the threshold to 7. At this threshold, all the word groups corresponding to the eight problem areas of the DGRST sub-file were found on the general proximity graph within a tightly meshed network of words.

This method of progressively lowering the threshold allowed us to identify the position and the relative weight of each DGRST problem area in an industrial/university network of problematisation. Five clusters (1, 2, 4, 5 and 8) corresponded to word associations which also defined specific problem areas in the general network at a relatively high threshold ($\geqslant$ 15). The remaining DGRST problem areas (clusters 3, 6, 7) appeared to be more original when compared with the core associations of the general network. At first sight it might be thought that they played a strategic role in creating new and previously unformed links in existing translation chains. But in fact two interpretations are possible. The problem areas might serve to create relationships between research themes generally considered as being independent of one another, or to strengthen those which already exist. In this case, the translation chains would no doubt be more precise and probably more efficient. But, on the other hand, and more prosaically, this action might correspond to the creation of 'neutralised' research themes which would allow the actors taking part in the Concerted Action Programme to work together on subjects of general, long-term interest without losing sight of their immediate scientific and/or market strategies. In the first case, the role of the DGRST would be strongly interventionist; in the second, it would be limited to obtaining provisional and tactical changes in actor strategies. A partial answer to this question will be found in the third section of this chapter.

Figure 9.2 represents a simplified version of the general network obtained at the 7 threshold. The positions of the eight DGRST word clusters (shaded areas) are shown together with a limited number of other problem areas. These shaded triangles are numbered 1 to 8 in accordance with the list of these clusters given above. The passageways between the various problem areas are indicated by a series of 'signposts' that were progressively identified as being the most important connecting terms on the basis of the proximity coefficient analysis carried out on the general file at the 30 and 15 thresholds. The arrows indicate links to words which are not shown on the map.

Two remarks can be made about Figure 9.2. The first concerns the

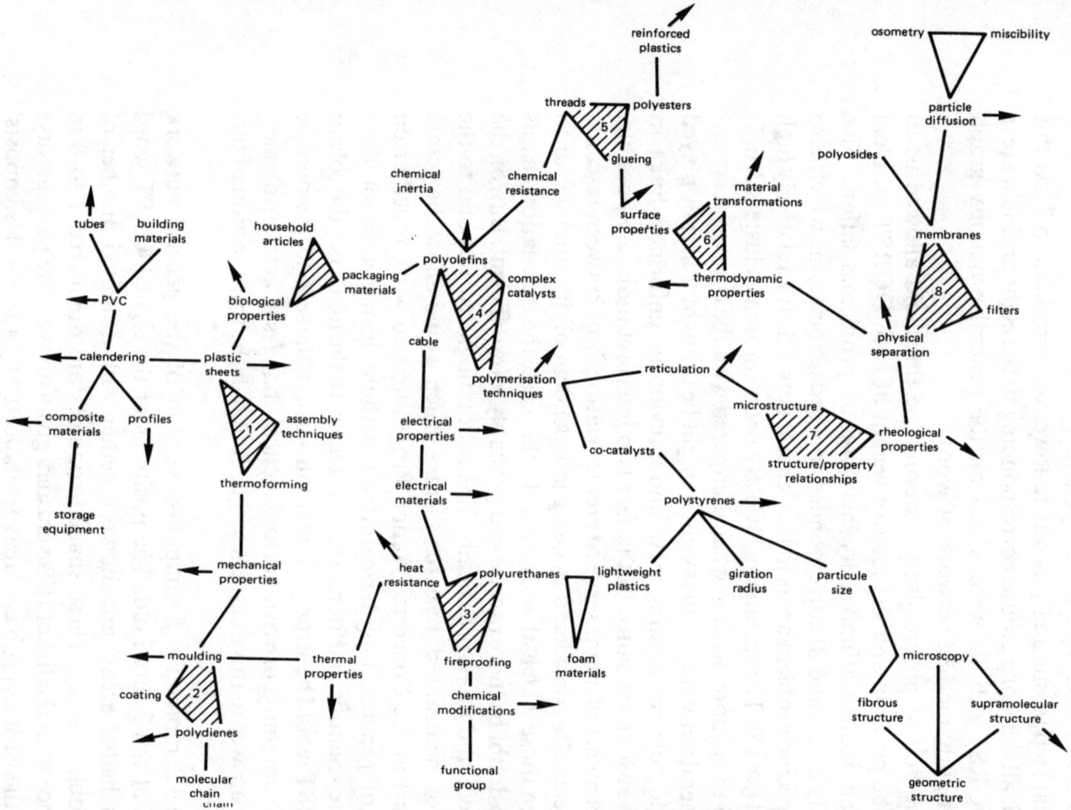

FIGURE 9.2 The position of DGRST problem areas in the general industry/university network of problematisation. Shaded areas indicate DGRST problem areas; arrows show links to words not shown on the map

role that different properties play in connecting the industrial processes identified on the maps. The only connection which is not linked in one way or another to the study of a particular property is the passage leading from cluster 3 to cluster 4 via the chain of lightweight plastics, polystyrenes and co-catalysts. This empirical evidence showing the strategic role played by polymer properties in linking together different areas of research serves *a posteriori* to justify the Committee's continued interest in promoting work to improve understanding of these properties (cf. Figure 9.1).

The second concerns the fact that each industrial process identified on the maps appears as a starting point which can lead to a specific research question or to a specific product. Take, for example, the problem of fireproofing polyurethanes (problem 3 on the map). The polymer points to the industrial implications of the research under way for the lightweight plastic market. The technique itself raises questions about the chemical modifications of the functional groups in the molecular structure of the substance which is fireproofed. Studying this technique in connection with new heat resistant materials has implications for the electrical industry. It must be remembered that this observation was made possible by positioning the problems raised within the framework of the DGRST Concerted Action Programme in a wider industry/university context. This fact would seem to suggest that the laboratories bidding for contracts from the DGRST spontaneously organised their interactions around industrial processes because, in this way, they could independently pursue the market and/or scientific strategies they judged appropriate.

Unfortunately, the summaries of proposals submitted to the DGRST do not give sufficient detail of individual proposals to make it possible to pursue this question further.

WHO IS INTERESTED IN THE PROBLEMS FORMULATED IN THE PROPOSALS SUBMITTED TO THE DGRST?

Table 9.1 provides an answer to the above question. Using the key-word definition of these problems which is indicated by means of the shaded triangular relationships on Figure 9.2, we automatically extracted all the documents in the total data file which contained at least one of the defining terms. We next classified these documents according to their interest in the various problems, the data sources analysed (DGRST, *Journal of Polymer Science, Plastiques Modernes et Elastomères*) and, in

TABLE 9.1 *The distribution of interests identified in the proposals submitted to the DGRST (percentages calculated using the column totals)*

		DGRST						
		Ind. (1)	*Univ.* (2)	*Pub. Res.* (3)	*Total DGRST* (4) (1)+(2)+(3)	*J. of Poly. Science* (5)	*Plas. Mod. & Elast.* (6)	*Total* (4) + (5) + (6)
Thermoforming Plastic Sheets	(no.)	23	3	8	34	184	250	468
	(%)	13	1.5	5	6.4	11.3	17.4	
Coating and Moulding Techniques	(no.)	10	12	14	36	116	562	714
	(%)	5.6	6.11	8.8	6.7	7.1	39	
Fireproofing Polyurethanes	(no.)	12	5	7	24	56	65	145
	(%)	6.7	2.5	4.4	4.5	3.4	4.5	
Polymerisation Techniques	(no.)	36	49	24	109	721	294	1125
	(%)	20.2	25	15	20.4	44.2	20.5	
Reinforced Polyesters	(no.)	9	5	11	25	60	199	284
	(%)	5.05	2.6	6.9	4.7	3.7	13.8	
Material Transformation	(no.)	20	18	37	75	174	23	272
	(%)	11.23	9.2	23.1	14	10.7	1.6	
Structure/Property Relationships	(no.)	15	17	17	49	107	23	179
	(%)	8.4	8.7	10.6	9.2	6.6	1.6	
Filtering Techniques	(no.)	53	87	42	182	212	21	415
	(%)	29.8	44.4	26.2	34.1	13	1.5	
Totals	(no.)	178	196	160	534	1630	1437	
	(%)	100	100	100	100	100	100	

the case of the DGRST, the industrial, university or public research laboratory making a bid for contract money.

Two types of information are thus found in Table 9.1. The first three columns of the table show the respective interests of industry, university and public research laboratories in each of the problems identified in the DGRST context, given their requests for funding. The total number of DGRST documents relating to a problem is given in column 4. This number can be compared with the total number of documents published on the same subject in the *Journal of Polymer Science* (column 5) and *Plastiques Modernes et Elastomères* (column 6). Columns 5 and 6 enable us to measure the interest that laboratories working in a scientific or industrial context have in the problems raised in the DGRST framework. The comparisons are facilitated by the percentages given with the document counts which are calculated using the column totals.

The information contained in the first three columns of Table 9.1 supports the hypothesis that the laboratories bidding for DGRST contracts spontaneously developed their applications in a way that made it possible for them to 'divide up the cake'. Industry seems to have a monopoly in the study of thermoforming plastic sheets, and to a lesser extent, the fireproofing of polyurethanes. The chosen areas of university activity are in polymerisation and filtering techniques. Public research laboratories are particularly interested in the study of material transformation. Moulding and coating techniques, reinforced polyesters and studies of structure/property relationships are of about equal interest to the three institutional partners.

It was perhaps necessary to divide the limited resources of the DGRST in a defined manner in order to help French science and industry stay internationally competitive. This hypothesis assumes that the problems which served to divide up the cake were also of interest in terms of what was taking place elsewhere. Clearly, this is the case. Columns 5 and 6 of Table 9.1 indicate that a sufficiently large number of documents were published in the *Journal of Polymer Science* and *Plastiques Modernes et Elastomères* to confirm the results of our proximity coefficient analyses: 1630 documents in the *Journal of Polymer Science* (80 per cent of the documents analysed) and 1437 industrial announcements in *Plastiques Modernes et Elastomères* (87 per cent of the analysed announcements) revealed an interest in the problems formulated in the projects submitted to the DGRST. The problems identified can quite properly be listed among the main elements that structure the industry/university network.

What was the degree of similarity between the centres of interest in the DGRST proposals and those found in our two other documentary sources? If interest in each problem was the same, the percentages given in columns 4, 5 and 6 of Table 9.1 would all be 12.5 per cent. In fact they are not, and in Table 9.2 we have used a plus or minus sign to indicate a positive or negative variation of 5 per cent (or less) from this control figure. It is easy to see that the dominant centres of interest in the DGRST context coincide much more closely with those found in the *Journal of Polymer Science* than those found in *Plastiques Modernes et Elastomères*. All the signs for the *Journal of Polymer Science* are oriented in the same direction except those that measure interest in the study of the way in which thermodynamic and surface properties affect material transformations. In the case of *Plastiques Modernes et Elastomères*, the opposite is generally true.

TABLE 9.2 *Dominant centres of interest in three documentary sources*

	DGRST	*Journal of Polymer Science*	*Plastiques Modernes et Elastomères*
1 Thermoforming plastic sheets	– –	–	+
2 Coating and moulding	– –	– –	+ + + + + +
3 Fireproofing polyurethanes	– –	– –	– –
4 Polymerisation techniques	+ +	+ + + + + +	+ +
5 Reinforced polyesters	– –	– –	+
6 Material transformation	+	–	– – –
7 Structure/property relationships	–	– –	– – –
8 Filtering techniques	+ + + + +	+	– – –

Plus or minus signs indicate a positive or negative variation of 5 (or less) from the value expected if interest for each problem was constant in the different documentary sources.

We can now use Tables 9.1 and 9.2 to comment on some of the questions which were raised in the preceding section when we looked at the way in which the proposals submitted to the DGRST fitted into the general network of problematisation.

We identified three problem areas which might eventually be important in the creation of new links in the translation chains (clusters 3, 6, 7). Two of these seem to be areas of sustained industry/university participation within the context of the DGRST Concerted Action Programme. They are the microstructural study of structure/property relationships (cluster 7) on the one hand, and the study of thermodynamics and surface properties in connection with material transformations (cluster 6) on the other. It is interesting to note that in both cases, public research laboratories show the greatest interest in the problems raised. This might indicate that these laboratories serve as a junction between industry and the universities in France. However, given the lack of enthusiasm for these two subjects shown in the other two documentary sources and, notably in *Plastiques Modernes et Elastomères*, it would seem that this aspect of the concerted action is limited to areas where major scientific and industrial interests are not at stake. With respect to the third problem – the study of the heat resistance of polyurethanes in connection with fireproofing techniques (cluster 3) – it would appear from the low level of interest in all three documentary sources that this area is not one that could serve to promote closer ties between industry and research. In the context of the Concerted Action Programme it is of particular interest to industry. Three contract proposals in this area were submitted to the DGRST by Ugine Kuhlmann and three more by the *Société Nationale de Poudres et d'Explosifs*. Both of these firms frequently sought grants from the DGRST. On the other hand, six small or medium-sized business firms presented requests in the area. They were no doubt hoping to resolve a specific problem encountered in their activity, because there is no evidence that they renewed their efforts to get research funds from the DGRST for other projects.

French universities generally show about the same general interest in the identified set of problems as laboratories publishing articles in the colloquium section of the *Journal of Polymer Science*. Two notable exceptions concern thermoforming plastic sheets and filtering techniques. Given the number of articles published on the question in the *Journal of Polymer Science*, the very low interest of French universities in the first problem area apparently has nothing to do with an *a priori* distinction between fundamental and applied research. It would seem to be best explained by the claim made above that thermoforming plastic sheets is a reserved field of action for industry in France.

French universities appear as a driving force behind what seems to be a very specific national preoccupation with filtering techniques (cluster

8). Table 9.1 shows that there are very few announcements of industrial applications using these techniques, but that French industry gives considerable attention to the area. As the low score for this theme in *Plastiques Modernes et Elastomères* suggests, this is probably a long-term interest which is considered promising in both industrial and scientific circles: cooperation can no doubt be envisaged because the secrecy clauses which generally apply in market-oriented research can, in this case, be avoided.

A final point can be made about the apparent lack of industrial interest in moulding and coating techniques and reinforced polyesters in the context of the Concerted Action Programme which can be contrasted with the large number of industrial announcements about these two subjects in *Plastiques Modernes et Elastomères*. No doubt French firms were unwilling to reveal their projects in two areas where vital economic and commercial interests were at stake.

TRANSLATION STRATEGIES

The word association patterns which appear on the maps generated by the Leximappe programmes suggest an approach which can be used to describe translation strategies with respect to the DGRST. Two types of links exist between problems. Those which appear on a proximity map indicate pathways which lead from one problem area to another. However, those which appear on an inclusion coefficient map indicate the way a problem relates to a variety of very general considerations. This latter type of relationship is evidence of a research community's effort to translate complicated scientific and technical concerns into a language which is more readily understood by non-specialists (see Chapter 7).

In the second section of this chapter, we explained how we progressively and selectively identified the main 'signposts' serving to indicate the principal passageways leading from one DGRST problem area to another in the macromolecular chemistry network of problematisation. Our inclusion coefficient analysis was carried out at a 0.470 threshold on the entire data file containing the summaries of proposals submitted to the DGRST, the articles from the *Journal of Polymer Science* and the industrial announcements in *Plastiques Modernes et Elastomères*. Three 'central poles' were found which focused and brought together a wide variety of scientific and technical interests: 'material composition', 'material structure' and 'material functions'. The different signposts and central poles identified were then used to classify the documents responsible for the word associations defining the DGRST problem areas into two categories.

The first category contained all the proposals submitted to the DGRST which concerned a specific problem area but which contained none of the terms serving to indicate the signposts and central poles of the global file. This indicates the lack of a translation strategy on the part of the actors submitting their proposals. No apparent attempt is made to justify an interest in a problem area by reference to general research themes or other research problems. We have already seen an example of this type of situation when we identified a group of small or medium-sized business firms that made proposals designed to solve specific problems encountered in connection with fireproofing polyurethanes.

The second category of proposals were those that related their research interests to other research problems or more general research objectives. Given that each DGRST problem area is identified by a set of key-words (see the shaded areas on Figure 9.2), the Multipol Leximappe programme (see Chapter 11) can be used to count the number of times that these key-words appear in the summaries of the proposals submitted to the DGRST in association with the words serving to define the signposts and the central poles of the global file. For example, Figure 9.3 can be used to study the outcome of translation strategies designed to link an interest in a specific problem area with a more general research theme. The eight problems formulated in the summaries of the proposals submitted to the DGRST were connected to one of the central poles when at least five proposals in the DGRST sub-file explicitly established this relationship.

Figure 9.3 shows that three problems – moulding and coating techniques, reinforced polyesters and fireproofing techniques are not linked to any of the central poles. Table 9.1 showed that these problems are of specific interest to industry. Coating and moulding techniques and reinforced polyesters are over-represented in *Plastiques Modernes et Elastomères* while fireproofing techniques, which receives relatively little attention, obtains its best scores in *Plastiques Modernes et Elastomères* and the proposals submitted by industry to the DGRST. On the other hand, the problems which are of particular interest to university and public research laboratories are all clearly linked in Figure 9.3 with the general themes of the field.

Table 9.3 identifies the industry, university and public research laboratories which submitted proposals to the DGRST containing the links between the problems and the central poles shown on Figure 9.3.

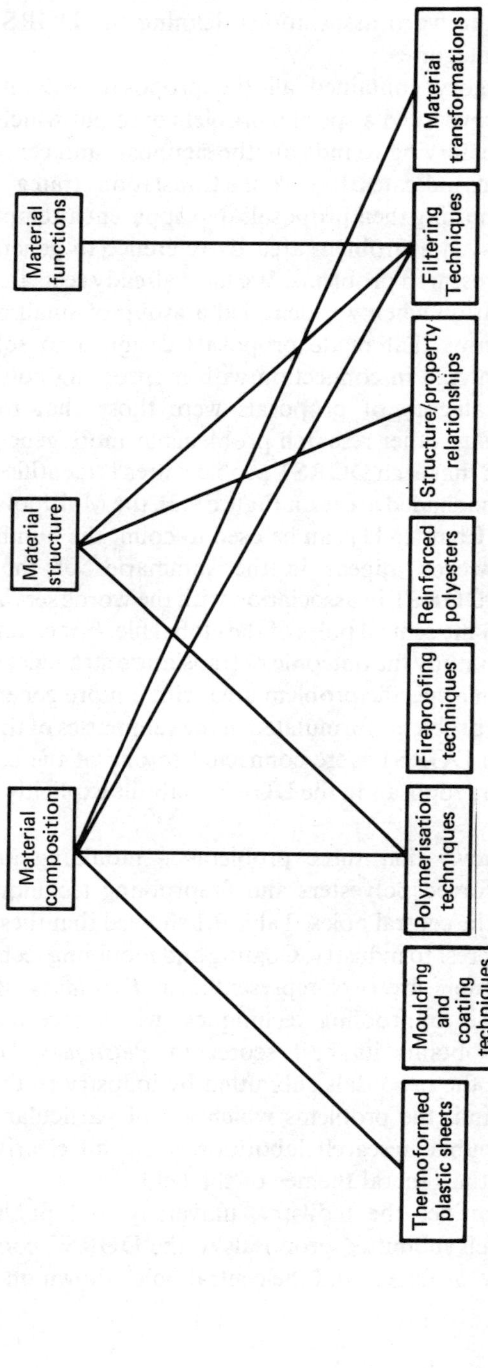

FIGURE 9.3 Links between problems and central poles

The percentage figures in Table 9.3 were obtained by dividing the number of documents linking a problem to a central pole by the number of documents concerning the problems given in columns 1, 2 and 3 of Table 9.1.

TABLE 9.3 *The role played by different partners of the DGRST Concerted Action Programme in positioning research problems with respect to general socio-economic objectives*

	Industry		*University*		*Public Research*	
	(no.)	(%)	(no.)	(%)	(no.)	(%)
Material composition						
thermoforming plastic sheets	12	**52**	0	0	6	**75**
polymerisation techniques	11	**30**	9	18	14	**58**
filtering techniques	17	**32**	16	18	13	**30**
material transformation	9	**45**	4	22	19	**51**
Material structure						
polymerisation techniques	11	**30.5**	9	18	5	**21**
filtering techniques	13	**24.5**	12	14	6	**14**
structure/property relationships	4	**27**	6	35	8	**47**
Material function						
filtering techniques	6	**11**	9	10	7	**17**

Percentages were obtained by dividing the number of documents linking a problem to a central pole by the number of documents concerning the problem which is given in columns 1, 2 and 3 of Table 9.1.

Table 9.3 reveals a paradox: university laboratories appear to be specifically interested in problems which are linked to central poles, but these links are essentially created by industry and, to a lesser extent, by public research laboratories. When this finding is combined with the results previously discussed in connection with Figure 9.3, the following interpretation can be suggested.

Both industry and university laboratories can retain certain problems for investigation without having to justify their choice by an explicit translation strategy. It would appear, however, that university laboratories rarely ventured outside their reserved domains of activity. By contrast, industrial laboratories often presented proposals in research areas which did not necessarily coincide with what one might expect to be their particular centres of interest. When they did so, they took care to justify their action with an explicit translation strategy linking their project proposal to themes of general long-term interest. This might be seen as an example of pure opportunism designed to get more money for research, but it might also be seen as the mark of a strategy designed to occupy the terrain. For example, when one looks in detail at the proposals presented by a large French chemical firm, Pechiney–St Gobain, the second of these two eventualities seems to apply. This firm submitted proposals to study each of the problems subordinated to the pole 'material composition', and carefully established the relationship with the general theme in each case. Its strategy appears to have consisted in positioning itself as a gatekeeper for research on ways of conferring physical qualities on plastic substances to make these suitable for various market applications.

We next considered the translation strategies linking together specific problems. The Multipol programmes allowed us to compare the action of universities and industry. We found that university interest in specific problems was often doubled by a concern to control the passageways leading to other research areas. This was not generally the case for industry.

CONCLUSION

The results described above concern the DGRST Concerted Action Programme for macromolecular chemistry in France. The data used covered the period from 1963 to 1975, but we have not retraced changes in the structure of industry and university research over the whole of this period because the complexity of such a study would not have allowed a brief presentation. Furthermore, the results we have presented in this chapter constitute a follow-up to a previous study published in 1979 which analysed the strategic relationships between DGRST and its contractants in more detail. Our concern in this chapter has been to show the utility of co-word analysis in evaluating some aspects of the cumulative impact of this science policy. Our results are still fairly general because we have been forced to aggregate the data in order to overcome problems encountered with our documentary sources. Nevertheless, we have been able to identify some of the more stable features of the research carried out in the context of the Concerted Action Programme.

We have used the proximity coefficient in order to identify the problems that spontaneously served to structure industry/university research given the proposals submitted to the DGRST. These problems occupied different positions within the general network of problematisation. Three problems seem to have played a role in linking together different parts of the global network; another problem (filtering techniques) has been shown to be of particular importance in the French context; the other problems coincided with some of the major problem areas appearing in the *Journal of Polymer Science* and *Plastiques Modernes et Elastomères*. We have thus identified signs of orthodoxy and heterodoxy in the proposals submitted to the DGRST.

We have also found that the majority of the DGRST problem areas concerned industrial processes. Why did university, industry and public research laboratories spontaneously organise their interactions around this set of problems rather than around another, perhaps equally suitable, set of questions? It is impossible to answer this question on the basis of a co-word analysis. A series of interviews would be required. However, use of the technique suggests several hypotheses. Study of the general network of problematisation suggests that the contractants could use industrial processes as common starting points in the definition of independent scientific and/or market strategies. Our results suggest both that individual institutions were left the responsibility of building bridges between fundamental and industrial research given their specific market and/or research strategies and that these strategies were not open for discussion at the Steering Committee level.

We have also found that the problems formulated in the DGRST proposals were used as a basis for dividing the limited resources of the Concerted Action Programme between the different partners bidding for contract money. But at the same time, they coincided with the centres of interest of laboratories publishing in the *Journal of Polymer Science* and the professional journal *Plastiques Modernes et Elastomères*. While the problems raised potentially allowed the French polymer community to stay competitive, the scientific literature was apparently used to a greater extent than the industrial literature as a means for determining appropriate centres of interest for French activity. This is yet another indication of a tendency to avoid questions with direct market implications.

Our study of translation strategies has revealed that university laboratories generally sought money to undertake work in problem areas where the legitimacy of their action was not open to doubt. They were more concerned with opening up passageways to other problem areas than in positioning their problems with respect to general research objectives in the field. Industry, and to a lesser extent, public research laboratories, appeared, on the other hand, to use the Concerted Action Programme as a means of extending their action to include areas which were not necessarily of immediate interest to them. When they did this, they took care to subordinate their proposals to long-term objectives of general interest to the polymer community. However, this can also be interpreted as an example of their effort to position themselves in a way which would allow them to reap the benefits of university research.

We are now in possession of a certain number of converging indicators which can be used to give a general interpretation of the impact of the DGRST science policy in the field of macromolecular chemistry. It would appear that the Concerted Action Programme was considered as a means of developing what Schumpeter (1926) called a 'pool of knowledge' about the mechanisms involved in the production and use of plastics. This interpretation would seem to reconcile the DGRST's efforts to restructure industry/university research with the general refusal of the participating partners to negotiate their individual scientific and/or market strategies.

NOTES

1. The words chosen by the Strasbourg students to index each document had necessarily to appear in the controlled vocabulary of the *Thesaurus*. The position of each word within its hierarchical classification scheme was identified by means of a nine-character number code. For example, the keyword 'chlorosulfone ethylene polymer' was coded 38 099 02 01, indicating that this polymer was found with the 'polyethylenes' (38 099 02 00) in the chapter on 'polyolefins' (38 099 00 00) in the section of the *Thesaurus* dealing with 'polymer families' (38 000 00 00). We found that a three level aggregation of our data was needed in order to identify the links which existed between industrial innovation and scientific research and in order to study the role played by the DGRST in creating these links. In other words, we only worked with the chapters of the *Thesaurus* defined by the first five characters of our number codes.
2. The notion of a 'simplified network' as a point in another network is developed in Chapter 2.

10 Pinpointing Industrial Invention: An Exploration of Quantitative Methods for the Analysis of Patents

MICHEL CALLON

INTRODUCTION

In the course of animated debates, specialists have slowly come to recognise that the study of patents can provide a limited but irreplaceable contribution to the knowledge of technical change.[1] Our aim in this chapter is not to discuss the general worth of this sort of study but rather to explore the use of co-word analysis to improve the study of the contents of a large number of patents.

The database used for this chapter is a file put together at our request by the INPI.[2] It contains 450 patents in the field of biotechnology which were filed with the European Patent Office in 1979, 1980 and 1981. For linguistic reasons, only those documents written in French or English were retained for indexing. Once those written in German were excluded, the file contained 268 patents distributed in the following manner:

Year	*1979*	*1980*	*1981*
Number of patents	53	106	109

The patents were indexed by the CDST. It was decided, for practical reasons, to limit the scope of indexation to the title, the summary and, to a lesser degree, the claims. This strategy was used for the whole file. Attention should be drawn to the fact that, given the experimental nature of this study, the indexation was performed with great care.

Two sub-files were then constituted for 1980 and 1981. The comparative analysis which we will undertake thus concerns two periods which have approximately the same number of documents. The aim is not to give a full analysis of the evolution of biotechnologies, but rather to apply the method of co-word analysis to the patents and giving an interpretation of the maps.

INCLUSION MAPS 1980, 1981

The first step in the study is to establish a set of inclusion maps. The goal of these maps is to reveal the central poles and the mediator words.

Once the maps are drawn, we are faced with a preliminary task of interpretation. Certain words unify the whole of the field without really adding new information. If they are deleted, the structure of the maps is simplified but not altered. This is the case in both 1980 and 1981 for the words 'production' and 'preparation'. These words occur so often because that is the terminology required by patent law. They index a large majority of the patents and can be put aside without inconveniencing this study.

Once this preliminary operation is completed, a series of central poles becomes evident. For 1980 these are the words: 'enzyme', 'bacteria', 'fermentation', 'antibiotic', 'bioreactor', 'food/feeding', 'anaerobic digestion', 'dosage form' and 'animal' (Figure 10.1). This last word is quite surprising, but the inspection of the words aggregated under this pole shows that it can be expanded to read: 'animal cell culture of vaccines'. In order to avoid confusion we will refer to this pole as 'animal tissue culture/vaccine'.

Both the central poles and the passageways are shown on the map (Figure 10.1). The low number of passageways (only two) reflect the very considerable fragmentation of the field. For the threshold used (0.333) a first and weak passageway links the pole 'bacteria' to the pole 'fermentation' by way of the mediator 'genetic engineering'. The fact that 'genetic engineering' only occurs three times suggests that a connection is only exceptionally established between these two poles.

The second passageway between 'fermentation' and 'antibiotics' is indicated by a long chain of words whose frequencies are all reasonably high and of about the same value. These words are shown on Figure

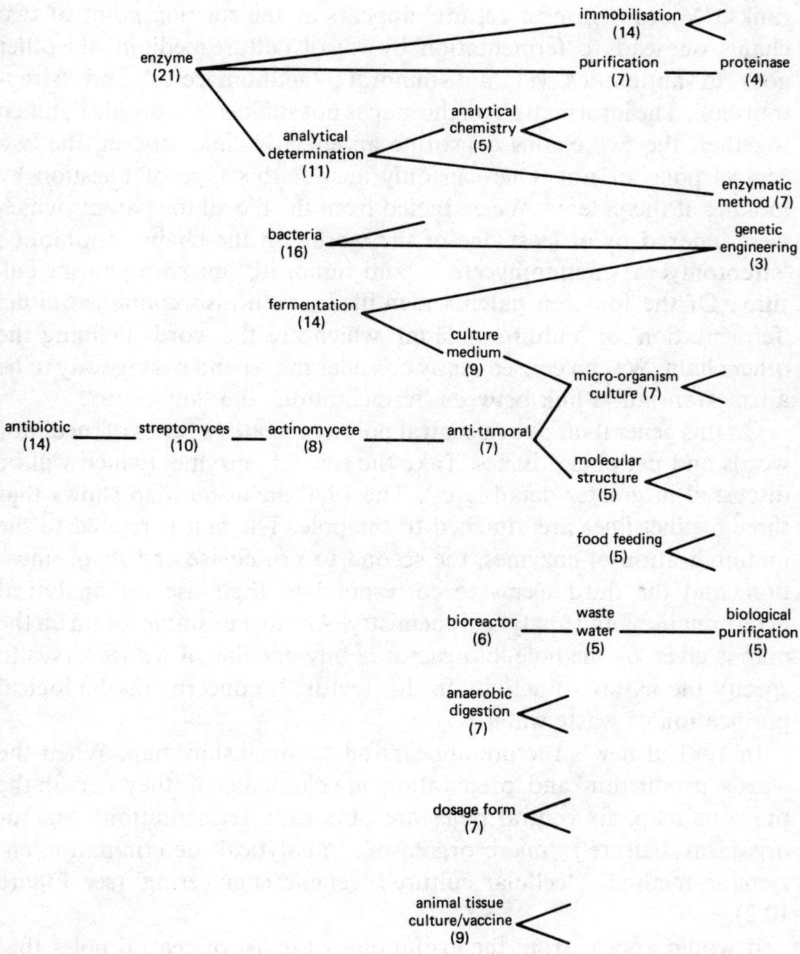

FIGURE 10.1 Inclusion map, 1980, central poles and passageways. Threshold 0.333. Frequencies in brackets. Some examples of mediator words and peripheral issues are given for 'enzyme' and 'bioreactor'

10.1. Their position with respect to the two central poles ('fermentation' and 'antibiotics') are indicated together with their frequencies and ranks. 'Micro-organism culture' appears as the starting point of two chains: one leads to 'fermentation' by way of 'culture medium', the other goes to antibiotics via 'anti-tumoral', 'actinomycetes' and 'streptomyces'. The information on the map is not sufficient to decide if, taken together, the two chains constitute an effective link between the two central poles or not. One can only answer this type of question by looking at the patents. We extracted from the file all the patents which were indexed by at least one of the words on the chain: 'antibiotic', 'streptomyces', 'actinomycetes', 'anti-tumoral', 'micro-organism culture'. Of the fourteen patents identified, seven also contained either 'fermentation' or 'culture medium' which are the words defining the other chain. We can consequently consider the second passageway to be a true translation link between 'fermentation' and 'antibiotic'.

On this general map each central pole aggregates a series of mediator words and peripheral issues. Take the case of 'enzyme' (which will be discussed in greater detail later). The 1980 inclusion map shows that three distinct lines are attached to the pole. The first is related to the immobilisation of enzymes, the second to proteinase and its purification, and the third seems to correspond to their use for analytical determinations and analytical chemistry. Another example found on the map is given by the pole 'bioreactor'. Only one line of words serves to specify the nature of activity in this sector. It concerns the biological purification of waste water.

In 1981, a new structure appears on the inclusion map. When the words 'production' and 'preparation' are eliminated as they were in the previous map, six central poles are obtained: 'fermentation', 'micro-organism culture', 'micro-organism', 'analytical determination/enzymatic method',[3] 'cellular culture', 'genetic engineering' (see Figure 10.2).

It would appear from the evolution of the list of central poles that there has been a considerable transformation of the maps over the two-year period. Only the pole 'fermentation' remains from one year to the next. But a more extensive examination of their structure will modify this initial impression.

The disappearance of the poles 'bacteria' and 'antibiotic' is more superficial than real. On the 1981 map, these two words are directly subordinated to 'fermentation' even though their frequencies are only slightly less than the frequency of this last word. Nevertheless, we did see on the 1980 map that no direct link existed between 'fermentation' and

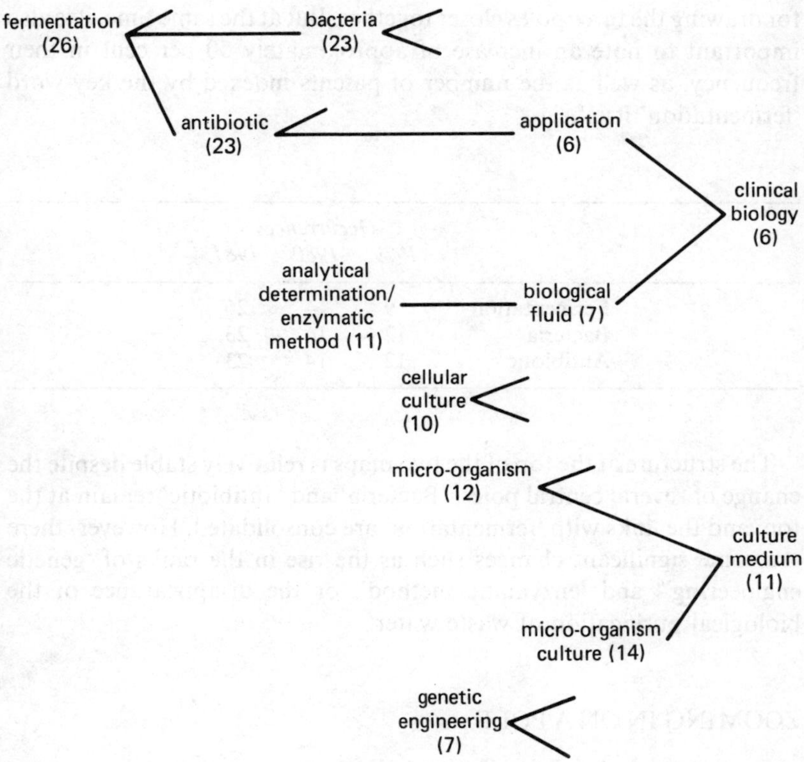

FIGURE 10.2 Inclusion map, 1981, central poles and passageways. Threshold 0.333

'bacteria'. With the 0.333 threshold, this means that less than five patents were simultaneously indexed using these words (we know that at least two documents associate these words because a link was shown on an inclusion map with a 0.100 threshold). In 1981 the relationship is stronger because the two words are found together in at least eight patents. What is true for the couple 'fermentation–bacteria' is also the case for the 'fermentation–antibiotic' pair. The figures are exactly the same: in 1980, no more than four patents can be simultaneously indexed using the two words; in 1981, they are found together in at least eight documents. The changes in the structure of the maps which result from the closer ties between 'fermentation', 'antibiotic' and 'bacteria' are to be expected, given the presence in 1980 of the passageways which we noted. No doubt the strengthening of these passageways is in fact responsible for drawing the three poles closer together. But at the same time, it is also important to note an increase of approximately 50 per cent in their frequency, as well as the number of patents indexed by the key-word 'fermentation' itself.

	Occurrences		
	1979	*1980*	*1981*
Fermentation	9	14	26
Bacteria	12	16	23
Antibiotic	12	14	23

The structure at the top of the two maps is relatively stable despite the change of several central poles. 'Bacteria' and 'antibiotic' remain at the top, and the links with 'fermentation' are consolidated. However, there are some significant changes such as the rise in the ranks of 'genetic engineering'[4] and 'enzymatic method', or the disappearance of the biological purification of waste water.

ZOOMING IN ON A POLE

To show how these changes can be analysed, we chose to follow the pole 'enzyme'. Its disappearance in 1981 seems quite mysterious. At the 0.333 threshold, this word does not appear on the 1981 inclusion maps although it still occurred sixteen times.

One way to find the reasons for this drastic change is by performing a zoom on the word 'enzyme'. Several zooming techniques are available. In this case we selected all the words of the inclusion maps which are linked directly or indirectly to 'enzyme'. If the initial threshold is not modified, the result obtained is a part of the general inclusion map. However, where the threshold is lowered, new lines appear. The word graph is much richer, but remains readable because it is only a segment of the total map. It can be interpreted like a normal inclusion map. Furthermore, the use of a very low threshold does not necessarily introduce noise into the analysis, because the word associations which are responsible for the links observed, no matter what their frequency, exist in the data file and, consequently, cannot necessarily be looked upon as probabilities which are not significant.

The result of the zoom carried out in 1980 on the word 'enzyme' is given in Figure 10.3. The thick dark lines show the associations found at a 0.333 threshold (they also appear on the global inclusion map) while the lighter lines indicate those which are found with an inclusion coefficient higher than 0.100.

The strong associations are: 'enzyme–analytical determination' (six patents), 'enzyme–purification' (four patents), 'enzyme–immobilisation' (nine patents). At the 0.100 threshold, three new relationships appear, introducing words in connection with 'enzyme', which did not

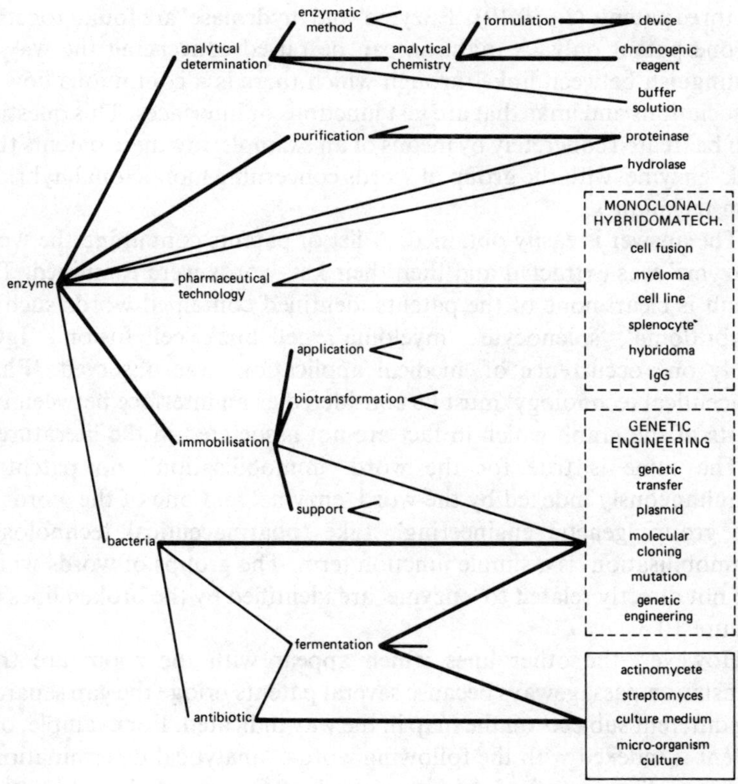

FIGURE 10.3 Zoom on 'enzyme', 1980. Thick line indicates 0.333 threshold; thin line indicates 0.100 threshold. Solid box encloses cluster of words strongly connected, where at least one patent exists which associates a word of the cluster with 'enzyme'. Broken box encloses words which are strongly connected where no word in the cluster is associated with 'enzyme' in any patent of the file

appear on the global inclusion map. Moreover, a group of words describing genetic engineering ('plasmid', 'molecular cloning', 'mutation'), is now associated with 'immobilisation' and 'bacteria'.

Let us first consider the new relationship which the zoom shows between 'enzyme' and 'pharmaceutical technology'. Of the eleven patents indexed by 'pharmaceutical technology', three also contain the word 'enzyme'. The inclusion coefficient consequently has a value of 0.27 (3/11). Had four patents been indexed by the two words, a line would have been drawn between them on the global map ($I = 4/11 = 0.36$).

The second relationship between 'enzyme' and 'bacteria' is produced by three patents ($I = 0.19$). 'Enzyme' and 'hydrolase' are found together in one patent only. A question can be raised concerning the way to distinguish between links through which there is a continuous flow of associations and links that are just junctions or interfaces. This question can be treated concretely by means of an example: are there patents that link 'enzyme' with the group of words concerning monoclonal hybridation techniques?

The answer is easily obtained. A list of patents containing the word 'enzyme' was extracted and then their key-words were examined. The result is clear: none of the patents identified contained words such as 'hybridoma', 'splenocyte', 'myeloma', 'cell line', 'cell fusion', 'IgG'. Only one occurrence of 'medical application' was observed. 'Pharmaceutical technology' must be considered as an interface between two parts of the graph which in fact are not associated in the literature.

The same is true for the word 'immobilisation': no patent is simultaneously indexed by the word 'enzyme' and one of the words in the group 'genetic engineering'. Like 'pharmaceutical technology', 'immobilisation' is a simple junction term. The groups of words which are not directly related to 'enzyme' are identified by the broken lines on Figure 10.3.

However, the other lines which appear with the zoom are true translation passageways because several patents bridge the gap separating different subjects on the map in the way indicated. For example, one patent is indexed with the following words: 'analytical determination', 'enzyme', 'fermentation', 'streptomyces', 'actinomycetes'; and another with: 'enzyme', 'pharmaceutical technology', 'antibiotic', 'biotransformation', 'bacteria', 'immobilisation'. This shows that the relationship between 'enzyme' and 'antibiotic production' is real.

Another point which should be noted is the fact that the patents responsible for the link between 'enzyme' and 'analytical determination' are also responsible for the links shown between 'analytical determination' and 'buffer solution', 'chromogenic reagent', 'analytical chemistry', and 'enzymatic method'. Furthermore, the patents which associate 'enzyme' and 'pharmaceutical technology' concern both the proteinase and its purification and antibiotic as well. On the other hand, no single patent directly links the words aggregated by 'analytical determination' with those associated with 'immobilisation' and the group of antibiotics.

This analysis could be continued in greater detail. However, all the observations converge: the zoom allows us to identify the translation links once the word group corresponding to monoclonal technology and genetic engineering are excluded from the graph. The word 'enzyme' lies at a pivotal position, linking together otherwise unconnected subjects. It is not a simple junction term. However, as we noted, a certain asymmetry exists: total integration of the different parts of the graph is not obtained because no patent associates 'enzymatic method' and 'actinomycetes', for example. A deeper analysis of the patents would reveal the importance of the link 'enzyme–analytical detemination' as a connecting element of the graph.

The comparison with the zoom carried out in 1981 on the word 'enzyme' is striking (Figure 10.4). As we said, 'enzyme', despite a frequency of sixteen, does not appear on the global inclusion map with the 0.333 threshold. This is clearly shown on the zoom where the thick dark lines are used to indicate inclusion coefficients superior to 0.333 and the lighter lines those which fall between 0.100 and 0.333. In 1981, the zoom shows that the word 'enzyme' aggregates three types of considerations. The first is identified by 'hydrolase', the second by a group of words organised around 'micro-organisms', the third by another group of words linked directly or indirectly to 'analytical determination' or 'enzymatic method'. In contrast with 1980, no patent establishes a direct link between these different considerations and only four patents link them to the words aggregating 'enzyme': 'antibiotics', 'bacteria' or 'fermentation'.

These observations confirm that the different parts of the graph are more independent in 1981 than in 1980. This, however, is not enough to prove that there has been a substantial change in the content of the patents from one year to the next. In order to take this question further, several observations can be made.

One of the essential features of the evolution which appears with the zooms is the reinforced position of 'enzymatic method'. This notion occurred seven times in 1980 and eleven times in 1981. However, its rank changed considerably. It moved from the position of a peripheral issue

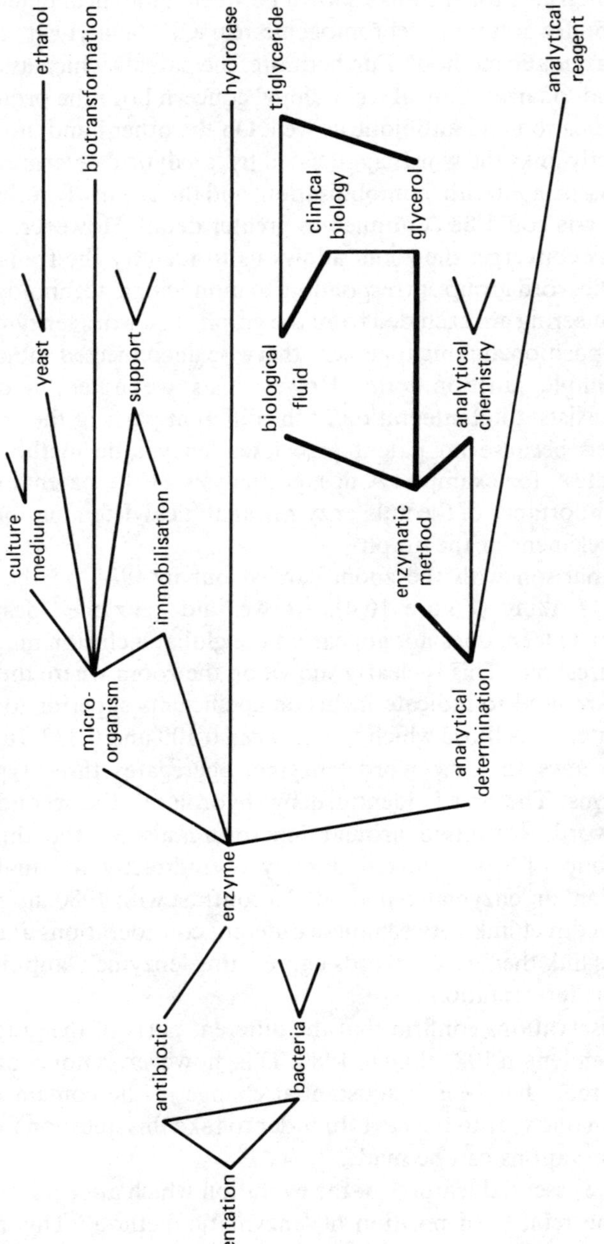

FIGURE 10.4 Zoom on 'enzyme', 1981. Thick line indicates 0.333 threshold; thin line indicates 0.100 threshold

to one of being a central pole focusing a wide range of interests. In 1980 'enzymatic method' was only linked to one word, 'analytical determination', to which it was subordinated at both the 0.333 and the 0.100 threshold (see Figure 10.3). This no doubt means that it served to group a very heterogeneous and dispersed set of techniques (because none of them appear on the zoom, despite a fairly high frequency of seven and a low threshold of 0.100) which have only one thing in common: the fact of being used to determine 'analytical determinations'.

In 1981, the situation has changed considerably: the context of 'enzymatic method' is much more restricted and precise because of the larger number of words which appear in connection with it. For example 'glycerol' and 'triglyceride' were totally absent from the key-word vocabulary in 1980 but appear in 1981 with a reasonably high frequency (five and three). 'Biological fluid' occurred four times in 1980, but did not appear on the zoom. In 1981, it is connected with 'enzymatic method' and has a frequency of seven. The same is true for 'clinical biology': absent from the maps in 1980 (frequency of two), it appears in 1981 linked to 'enzymatic method' and has a frequency of six.

Some final remarks can be made concerning the two zooms. The patents concerning 'proteinase' have practically disappeared (frequency passed from four to one). The same is true for the patents indexed by 'immobilisation' (the frequency passed from fourteen to eight). All links with 'pharmaceutical technology' have also disappeared, which was to be expected given our interpretation for 1980.

These observations can now be used to interpret the complete disappearance of 'enzyme' in 1981 after having been a central pole on the inclusion maps in 1980. We noted that at the 0.100 threshold, 'enzyme' once again appeared in a dominant position in 1981, even though its connecting power had weakened. It serves more as an interface between different fields than a bridge actually linking them together. In addition, a number of the notions which it had aggregated in 1980 are of less importance in 1981. In comparison, the now dominant position of 'enzymatic method' has served to redefine word relationships and introduce new notions.

This example shows how the method, through the simultaneous examination of the inclusion maps and the zooms, enables the researcher to rapidly identify some overall changes in the structure of patents in the field and to ascertain the main features of such reorganisation. As we will now show, this analysis can be taken further by using the proximity maps.

THE PROXIMITY MAPS

In 1980 and 1981, the proximity maps have a largely disconnected structure for the chosen threshold of 5. In both years they are constituted by a small number of totally connected components which are completely independent from one another (four the first year, three the second year). The content of these problem areas varies noticeably from one year to the next.

We will give a first example of interpretation of these maps which follows what has just been said about the evolution of the pole 'enzyme'.

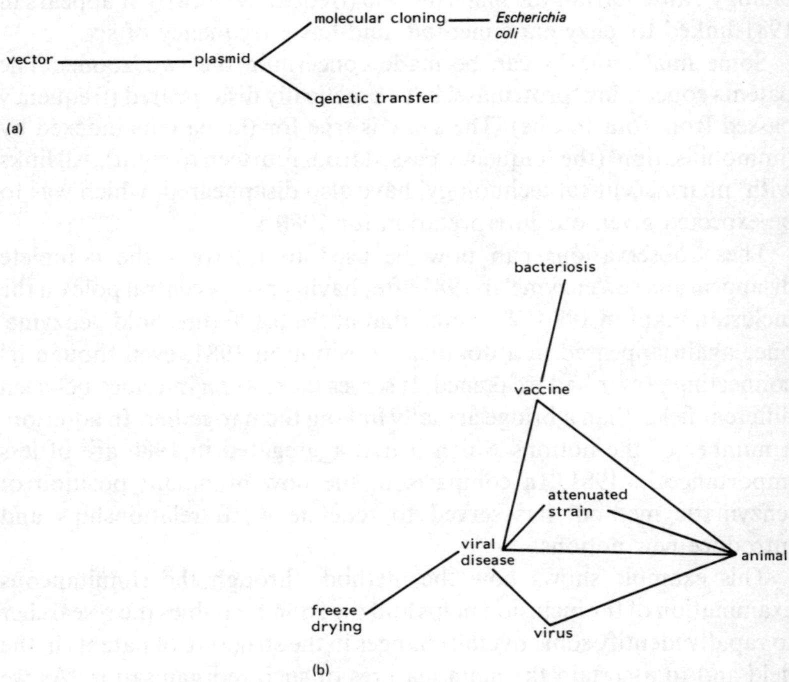

FIGURE 10.5 Proximity map, 1980 (a) first component, (b) second component, (c) third component, (d) fourth component

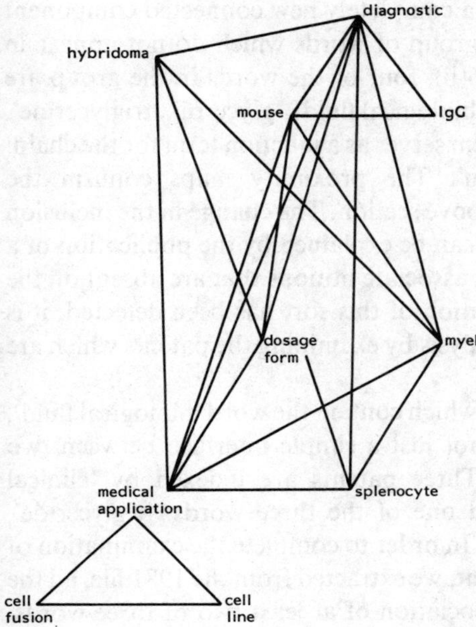
hybridoma
diagnostic
mouse
IgG
dosage form
myeloma
medical application
splenocyte
cell fusion
cell line

(c)

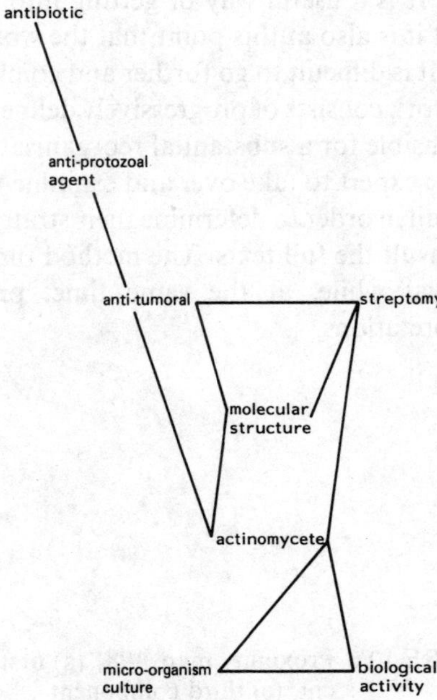
antibiotic
anti-protozoal agent
anti-tumoral
streptomyces
molecular structure
actinomycete
micro-organism culture
biological activity

(d)

It is interesting to note that a completely new connected component appears in 1981 containing a group of words which do not appear in 1980. As shown in Figure 10.6(b), four of the words in the group are closely linked to one another: 'biological fluid', 'glycerol', 'triglyceride', 'enzymatic method'. One of them serves as a junction term for the chain: 'clinical biology', 'application'. The proximity maps confirm the interpretation we gave in the above section. The change in the inclusion maps with respect to 'enzyme' can be explained by the publication of a small number of patents which associate notions that are absent on the 1980 maps. Once a transformation of this sort has been detected, it is possible to go further in the analysis by examining the patents which are responsible for it.

When we look at the patents which contain the word 'biological fluid', it is clear that this notion is not just a simple interface between two distinct parts of the graph. Three patents are indexed by 'clinical biology', 'biological fluid' and one of the three words 'triglyceride', 'glycerol', 'enzymatic method'. In order to complete the examination of patents linked to this component, we extracted from the 1981 file, all the patents responsible for the association of at least two of three words: 'glycerol', 'triglyceride' and 'enzymatic method'. Four patents were identified. Table 10.1 gives the list of words which were used to index them. It is a useful way of getting into patent content.

But it is also at this point that the work of the co-word analyst must stop. It is difficult to go further and comment on the meaning of the list. The work consists of progressively delineating a small number of patents responsible for a substantial reorganisation of the maps. It is now time for the expert to take over and examine the content of these documents in detail in order to determine their strategic importance. He would have to consult the full texts. The method functions as an aid for document retrieval while, at the same time, providing a number of initial interpretations.

FIGURE 10.6 Proximity map, 1981 (a) first component, (b) second component, (c) third component

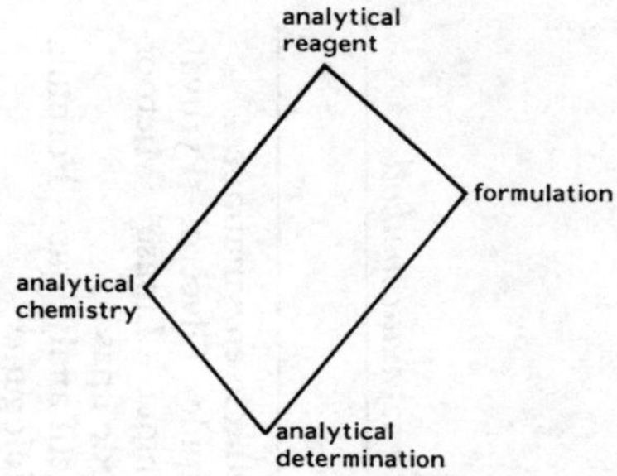
analytical
reagent
formulation
analytical
chemistry
analytical
determination

(a)

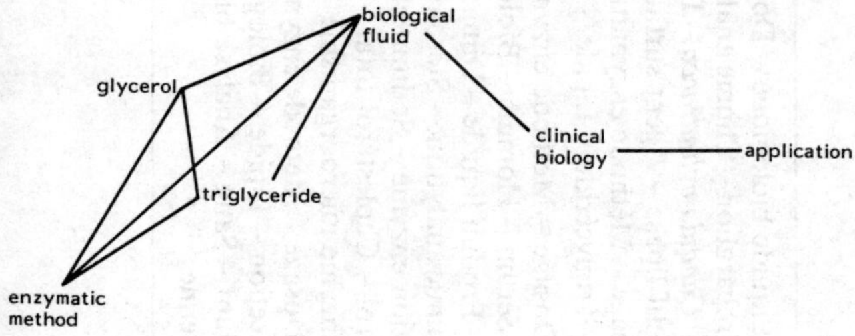
biological
fluid
glycerol
clinical
biology
application
triglyceride
enzymatic
method

(b)

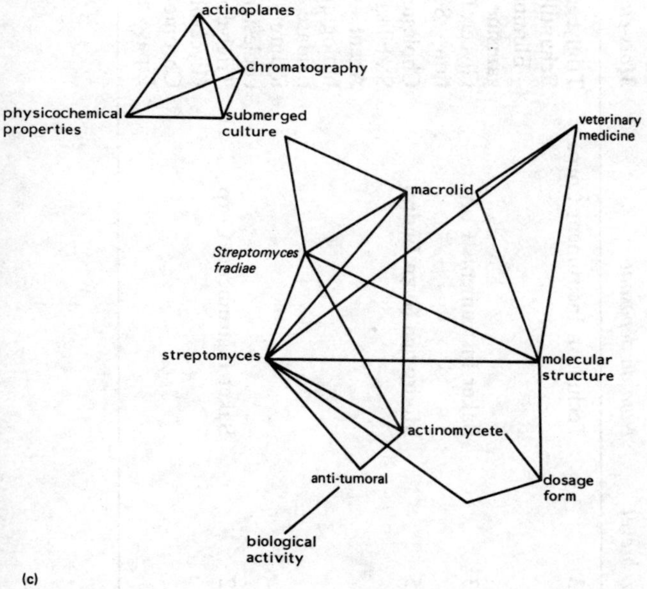
actinoplanes
chromatography
physicochemical
properties
submerged
culture
veterinary
medicine
macrolid
Streptomyces
fradiae
streptomyces
molecular
structure
actinomycete
anti-tumoral
dosage
form
biological
activity

(c)

TABLE 10.1 *Patents indexed in 1981 by at least two of the three words: 'glycerol', 'triglyceride', 'enzymatic method'*

No. brevet	*Nom du déposant*	*Mots-clés*
33	Technicon Instrument Corp.	Triglycéride – Liquide biologique – Dosage – Méthode enzymatique – Echantillon – Préparation – Chimie analytique – Lipide – Glycérol – Pyruvate – Elimination – *Candida cylindracea* – Biologie clinique – Lipase – Microorganisme – Réactif lipase – Agent surface – Méthode lipase
93	Baker Instruments Corp.	Glycérol – Dosage – Méthode enzymatique – Réactif analytique – Formulation – Sérum – Triglycéride – Lipide – Liquide biologique
185	Modrovich (Ivan Endre)	Cholestérol – Dosage – Méthode enzymatique – Réactif analytique – Stabilisation – Sérum – Homme – Biologie clinique – Solution tampon – Agent surface – Produit liquide – Diluant – Electrode oxygène – Kit – Liquide biologique – Chimie analytique – Stéroide – Cholestérol esterase – Cholestérol oxidase – Solution enzyme – Sodium cholate – Triton X-100 – Potassium iodure – Glycérol – Cholestérol total – Antipyrine (amino-4) – Kit analyse cholestérol – Enzyme microorganisme
212	Baker Instruments Corp.	Triglycéride – Dosage – Méthode enzymatique – Sérum – Homme – Animal – Glycérol – Détection – Lipide – Biologie clinique – *Pseudomonas* – Chromobactérium – Sang – Analyse biochimique – Liquide biologique – Lipase microbienne

The problem area which we have just studied was first identified in 1981. It is also interesting to look at areas which do not appear to have changed much between 1980 and 1981. One of these is identified by the following words (Figures 10.5(d) and 10.6(c)).

	Occurrences	
	1980	*1981*
biological activity	6	10
actinomycete	8	18
molecular structure	5	12
streptomyces	10	15
anti-tumoral	7	4

An immediate observation is the considerable change in frequencies, notably with respect to 'actinomycete'. In 1980 three words, 'actinomycete', 'streptomyces' and 'anti-tumoral', served as the starting point to a passageway relaying the general poles 'antibiotic' and 'fermentation'. The analysis showed that it was a true translation chain, a finding which is now confirmed by the proximity map. Four words ('actinomycete', 'streptomyces', 'anti-tumoral', 'molecular structure') are linked together in a two-by-two cluster. The importance of the role played by this translation chain in the dynamics of the field is underlined by the stability of the problem area that it organised, as well as by the clustering capacity of 'actinomycete' in both 1980 and 1981.

This stability should not lead us to neglect an important change: in 1981 a new word appears, 'submerged culture', which both extends the problem area and occupies a pivotal position. In 1980, the same word is not found in the file. By 1981, it has a frequency of eight. 'Submerged culture' also appears as a mediator word in the inclusion map for 1981 serving as a passageway between 'bacteria' and 'antibiotic'. Other words appear in connection with 'submerged culture': 'macrolid' (frequency zero in 1980, six in 1981) and 'actinoplanes' (respectively zero and three).

We extracted the seven patents indexed by 'submerged culture' in order to study the changes which were taking place (Table 10.2). All these patents contain the word 'antibiotic', but two distinct groups may be discerned. The first groups the patents linking 'submerged culture' to 'macrolid', 'streptomyces' and 'actinomycete'; the second contains the patents which associate 'actinoplanes', 'physicochemical properties', 'chromatography' and, of course, 'submerged culture', but without the words 'macrolid','streptomyces' or 'actinomycete'.

TABLE 10.2 *Patents indexed in 1981 by 'submerged culture'*

No. brevet	*Nom du déposant*	*Mots-clés*
First sub-family		
148	*Eli Lilly and Co.*	*Macrolide – Antibiotique – Préparation – Structure moléculaire – Fermentation – Production – Culture profonde – Mutation – Recombinaison – Souche – Streptomyces – Streptomyces fradiae* – Actinomycète – Tylactone
215	Eli Lilly and Co.	Antibiotique – Macrolide – Préparation – Médecine vétérinaire – Structure moléculaire – Production – Culture microorganisme – Culture profonde – Streptomyces – *Streptomyces fradiae* – Actinomycète
216	Eli Lilly and Co.	Antibiotique – Macrolide – Structure moléculaire – Production – Culture microorganisme – Culture profonde – Streptomyces – *Streptomyces fradiae* – Actinomycète – Tylosine – Mycarosyltylactone
244	Eli Lilly and Co.	Macrolide – Antibiotique – Production – Fermentation – Culture profonde – Forme pharmaceutique – Médecine vétérinaire – Structure moléculaire – Streptomyces – *Streptomyces fradiae* – Actinomycète

Second sub-family		
7	Gruppo Lepetit SPA	Antibiotique – Propriété physicochimique – Production – Actinoplanes – Culture profonde – Fermentation – Séparation – Chromatographie colonne – Bactérie – Antibiotique A/16686-A2 – Actinoplanes ATTC 3376
42	Eli Lilly and Co.	Antibiotique – Production – Actinoplanes – Fermentation – Culture profonde – Mélange complexe – Séparation – Propriété physicochimique – Technologie pharmaceutique – Facteur croissance – Chromatographie couche mince – Chromatographie phase liquide – Bactérie – Antibiotique A-4696 – *Actinoplanes missouriensis*
46	Ely Lilly and Co.	Antibiotique – Production – Actinoplanes – Mutation – Fermentation – Culture profonde – Purification – Chromatographie – Propriété physicochimique – Composition chimique – Facteur croissance – Aliment pour animal – Technologie pharmaceutique – Peptide – Bactérie – Antibiotique A-4696G – *Actinoplanes missouriensis*

TABLE 10.3 *Patents indexed in 1980 by at least one of the four words: 'actinomycete', 'streptomyces', 'molecular structure'* (structure moléculaire), *'anti-tumoral'* (anticancéreux)

No. brevit	*Nom du déposant*	*Mots-clés*
12	Kyowa Hakko Kogyo Co. Ltd	Urate oxidase – Production – Fermentation – Streptomyces – Dosage – Urique acide – Liquide biologique – Enzyme – Actinomycète
38	Takeda Yakuhinkoyo Kabushiki	Acaricide – Fongicide – Activité biologique – Production – Analogue – Préparation – Structure moléculaire – Fermentation – Végétal – Streptoverticillium – Pesticide – Actinomycète – Mildiomycine
60	UOP Inc.	Glucose phosphate isomerase – Purification – Immobilisation – Thermostable – Traitement thermique – Microorganisme – Support réaction – Alumine – Ethylèneimine polymère – Reticulation – Agent bifonctionnel – Precipitation chimique – Enzyme – Streptomyces – Actinoplanes – Glutaraldehyde
70	Sanraku Ocean Co. Ltd	Antibiotique – Anthracycline – Streptomyces – Anthracyclinones – Biotransformation – Mutation – Production – Composition chimique – *Streptomyces coeruleorubidus* – *Streptomyces peucetius* – Daunomycine – Baumycine
120	Yamanouchi Pharmaceutical Co. Ltd	Corynebacterium – Extrait cellulaire – Solution – Protéine – Glucide – Point isoélectrique – Anticancéreux – Antiallergique – Immunogencité – Bactérie
121	Sumitomo Chemical Co. Limited	Anticancéreux – *Acinetobacter calcoaceticus* – Immunostimulant – Culture cellulaire – Toxicité – Préparation – Isolement – Propriété chimique – Propriété physique – Traitement chimique – Propriété biologique – Test activité – Effet secondaire – Barium hydroxide
195	Eli Lilly and Co.	Préparation – Chimiothérapie – Hyperuricémie – Sporiasis glucoside – Glycosylation – Production – Antibiotique – Streptomyces – *Streptomyces aureofaciens* – *Streptomyces candidus* – Actinomycète – Mycophénolique acide dérivé – Goutte pathologie

219	Kyowa Hakko Kogyo Co. Ltd	Antibiotique – Activité biologique – Préparation – Technologie pharmaceutique – Culture microorganisme – Antibactérien – Anticancéreux – Propriété physicochimique – Production – Streptomyces – *Bacillus subtilis* – Actinomycète – Bactérie – *Streptomyces olivaceus* – Antibiotique DC-45-A – Antibiotique DC-45-B1 – Antiobiotique DC-45-B2 – Trioxacarcine A – Trioxacarcine B – Trioxacarcine C
232	Sanraku Ocean Co. Ltd	Antibiotique – Anthracyclines – Anticancéreux – Production – Fermentation – Structure moléculaire – Streptomyces – Actinomycète – Aclarubicine (Hydroxy-2) – Aclavinone (Hydroxy-2)
242	Sankyo Co. Ltd	Inhibiteur enzyme – Production – Culture microorganisme – Activité biologique – Application – Propriété physicochimique – Streptomyces – Actinomycète – Oriséolique acide – *Streptomyces griseoaurantiacus* 3′:5′ – Cyclic-Nucléotide phosphodiesterase
269	Takeda Yakuhin Kogyo Kabushiki Kaisha	Antibiotique – Structure moléculaire – Antiprotozoaire – Anticancéreux – Production – Culture microorganisme – Forme pharmaceutique – Streptomyces – Chainia – Streptosporangium – Actinomycète – Antibiotique C-15003 PHO
271	Glaxo Group Ltd	Antibiotique – Production – Fermentation – Souche – Charactérisation – Purification – Streptomyces – Actinomycète – Cephalosporine – Cephamycine
274	Takeda Yakuhin Kogyo Kabushiki Kaisha	Antibiotique – Production – Acylation – Antiprotozoaire – Anticancéreux – Activité biologique – Structure moléculaire – Culture microorganisme – Nocardia – Actinomycète – Antibiotique C-15003 PHM
276	Takeda Yakuhin Kogyo Kabushiki Kaisha	Antibiotique – Production – Anticancéreux – Antifongique – Antiprotozoaire – Fermentation – Structure moléculaire – Streptomyces – Chainia – Actinomycètes – Antibiotique C-15003 PND – Antibiotique C-15003 PND-Ansamitocine dérivé

Once again, examining the maps, or, more precisely, a relatively stable connected component of these maps, has allowed us to identify certain structural reorganisations and the documents which are responsible. The interpretations of the different maps so far suggested lead naturally to a discussion of different strategies used by actors when taking out their patents.

THE ACTORS INVOLVED

We will limit our discussion of actors' strategies to the previous examples. Two general remarks can be made. First, given the way in which the data file was constructed, no general conclusions can be drawn from this discussion. Once again, the aim of this chapter is to test the possibility of applying the method to a limited patent data file which is not necessarily representative, not to produce an exhaustive analysis of developments in the biotechnology field. Second, the interpretation which we will propose does not concern the actors themselves, but rather their patent strategies in Europe.

In what follows we will show that the map transformations constitute an appropriate starting point in order to identify participating actors and characterise their patenting strategies.

By studying the evolution of the word 'enzyme', we were able to identify the role played by a small number of patents defining a new problem area in 1981. Table 10.1 shows the coherence of the patents identified and also the variety of firms involved: Baker (two patents), Modrovich (two patents), Technicon Instrument Corporation (one patent). These patents, simultaneously taken out by three different firms, concern centres of interest which were not apparent in the 1980 data file.

We will now look at the problem area which remained stable over the period from 1980 to 1981 and which revolved around three terms: 'actinomycete', 'molecular structure', 'streptomyces'.

In 1980 these organising terms were associated with 'anti-tumoral'. The list of patents containing at least one of these four words is presented in Table 10.3. Fourteen patents in all are identified. The table shows that an overwhelming majority of companies involved in this field were Japanese (eleven of the fourteen). Of these fourteen patents, four contain only one of the four words used for extraction. Therefore, the remaining ten patents are responsible for this connected component. Of these ten patents, eight are Japanese (four from the Takeda Company).

In 1981 we saw that a new word ('submerged culture') was introduced which extended the problem area notably by the fact that it led to new associations corresponding to the definition of two distinct patent families (Table 10.2). This reorganisation was due to 'macrolid' and 'actinoplanes' as well as 'submerged culture', none of which appeared in 1980. The patents indexed by 'submerged culture' have already been indicated (Table 10.2). The three occurrences of 'actinoplanes' corresponds to three patents which define one of the connected sub-components of the problem area that retained our attention (Table 10.2, nos 7, 42 and 46). Two of these patents were deposited by Eli Lilly, the third by Gruppo Lepetit. Their content is described by the list of all the key-words they contained which is given in Table 10.4. In addition, six patents are indexed by the word 'macrolid': four patents were taken out by Eli Lilly and two others by Merck and Co. The four Eli Lilly patents were already present on Table 10.2, indexed by 'macrolid' and 'submerged culture' (patent nos 148, 215, 216, 244). The two Merck patents can be found in Table 10.5. Neither of them were indexed by 'submerged culture'.

We can now describe the patenting strategies of the identified firms. If we ignored Eli Lilly patents, we would have two distinct sub-fields, the word 'submerged culture' being, in this case, completely separated from the context 'macrolid', 'streptomyces', 'actinomycete'. If, on the other hand, we also ignored the two Merck patents, then 'macrolid' would disappear and the remaining connected component would be defined by 'anti-tumoral', 'actinomycete' and 'streptomyces'. Eight patents taken out by seven firms are responsible for this component: Hoffmann Laroche (2), Upjohn Company (1), Gruppo Lepetit (1), Fujisawa (1), Shronogi (1), Zardan (1) and Kowa (1).

TABLE 10.4 *List of key-words with frequencies in brackets*

Chromatographie colonne (1)	Technologie pharmaceutique (2)
Chromatographie couche mince (1)	Actinoplanes (3)
Chromatographie phase liquide (1)	Antibiotique (3)
Actinoplanes ATTC 3376 (1)	Bactérie (3)
Antibiotique A/16686 A2 (1)	Culture profonde (3)
Antibiotique A 46 96 G (2)	Fermentation (3)
Actinoplanes missouriensis (2)	Production (3)
Facteur croissance (2)	Propriété physicochimique (3)
Séparation (2)	

TABLE 10.5 *Key-word contents of Merck patents*

No. brevets	*Nom du déposant*	*Mots-clés*
158	Merck and Co. Inc.	Antiparasitaire – Parasitose – Animal – Médecine vétérinaire – Structure moléculaire – Activité biologique – Technologie pharmaceutique – Macrolide – Production – Streptomyces – Actinomycète – Antibiotique C-076 dérivé – *Streptomyces avermilitis*
236	Merck and Co. Inc.	Macrocycle – Lactone – Antibiotique – Production – Fermentation – Pesticide – Insecticide – Acaricide – Nematicide – Activité biologique – Streptomyces – Structure moléculaire – Macrolide – Actinomycète

In other words, in 1981, seven firms continue and consolidate a patenting strategy which was already used in 1980. One firm, Eli Lilly, seems to be developing an original activity in the area. It is working both to consolidate an already existing strategy and to extend its centres of interest to include a new set of associations organised around 'submerged culture'. Other forces of reconstruction in this area are Merck and Co. who are specifically interested in the link 'macrolid', 'streptomyces', 'actinomycete' and Gruppo Lepetit who is consolidating the association 'submerged culture', 'actinoplanes'. The patenting strategy of Eli Lilly is unique: it is the only firm to deposit patents which are directly connected to both the 'actinoplanes' family and the 'macrolid' family. It consequently seems legitimate to consider that the extension-reorganisation of this problem area between 1980 and 1981 is due to the action of Eli Lilly.

It is perhaps interesting to look now at a problem area which we have not yet discussed. Between 1980 and 1981 a connected component organised around the words: 'hybridoma', 'mouse', 'myeloma', 'cell fusion', 'IgG', 'cell line', disappears from the proximity maps. Given that the frequency of these words in 1980 is fairly high (around five) the fact that they totally disappear from the indexing vocabulary in 1981 is significant. This change is confirmed by the inclusion maps.

Five patents are responsible for the connected component identified in 1980. They were all deposited by Ortho Pharmaceutical Corp. and surprisingly we have no evidence that Ortho took out other patents than the five identified.[5] No other firm deposited a patent concerning monoclonal technologies.

CONCLUSION

This chapter has been devoted to an exploration of the co-word analysis as a method for studying the content of a patent database.

The limits of this type of database are now well known. Nevertheless, if one proceeds with prudence, patents have been shown to be a unique source of information about the mechanism of innovations and the dynamics of technological development.

The co-word analysis technique offers a means of describing these mechanisms and dynamics. The inclusion maps were used to identify the stable features and significant transformations which characterise the development of an activity. Passageways between a variety of centres of interest were also identified and their role in preparing changes in the structure of the maps were discussed.

We used zooms as a means of carrying out a detailed study of why 'enzyme' totally disappeared from the inclusion maps in 1981, after having been a central pole in 1980. This zooming technique showed that this abrupt change between the two periods was more than a simple fluctuation; it was linked to the appearance of a small number of patents which introduced new centres of interest and reorganised existing relationships. The patents in question were identified without difficulty.

The preceding analysis was both confirmed and carried further with the proximity maps. These maps allowed us to identify specific and independent problem areas for 1980 and 1981. For specific cases we showed how the proximity maps (a) completed the information given by the inclusion maps and (b) supplied us with the opportunity of determining and analysing the factors behind changes in the contents of a problem area.

As a final step in the study, we showed how the map of transformations could be explained in terms of the patent strategies of certain firms. Two cases were examined: one concerned a significant transformation in the structure of an inclusion map; the other the consolidation and reorganisation of a problem area identified on the proximity maps.

The results of our study appear to be positive. The changes in the map structures suggested questions; their study led us to identify a group of documents and actors whose role could be considered of strategic importance. Furthermore, the theoretical analysis outlined in Chapter 6 is confirmed. Because they define obligatory passageways, patents could be legitimately reduced to a series of associated words. Nevertheless the problem of patent indexation remains. The solution adopted for this chapter is too costly and consequently not entirely satisfactory. In Chapter 12, when we consider the future developments of the co-word analysis technique, we discuss programmes which are currently being set up for the semi-automatic indexation of full texts.

NOTES

1. See, for example, Freeman (1983), Schmookler (1966), Pavitt (1982), OECD (1980) and OECD (1982).
2. INPI (Institut National de la Propriété Industrielle) is the French National Patent Office.
3. The central pole is in fact 'analytical determination' but it aggregates 'enzymatic method' which has a similar frequency.
4. This evolution is confirmed by statistics supplied by the US Patent Office which show that for 'genetic engineering' 4 patents were delivered in 1979, 10 in 1980, 38 in 1981, 122 in 1982.
5. The Ortho Pharmaceutical Corp. was identified in the Patent Profiles elaborated by the US Patent Office as having deposited eight patents for monoclonals in 1982. It does not appear to be as active in the other fields covered by the Patent Office survey.

11 Technical Issues and Developments in Methodology

JEAN-PIERRE COURTIAL

In the preceding chapters, the emphasis has been on the presentation of the results of co-word analysis and the interpretation of the scientific and technological fields that were studied in this way. The method itself has been outlined and justified in Chapter 7, but there are a number of technical issues that deserve more detailed treatment, if only to satisfy the curiosity of the reader. In addition, the discussion of these issues allows us to point out some further developments in methodology.

To illustrate the technical and methodological points, we shall use data from the study of dietary fibre research; the main findings were published in Callon, Courtial, Turner and Bauin (1983), and have also been considered in Chapter 7. A selection of articles on dietary fibre, and related key-words (because this particular concept was not prominent at all in the first part of the period studied) was extracted from the PASCAL database of the French CNRS. The file contains 172 articles for the period 1973–8, and has a vocabulary of 144 key-words.

After justifying the choice of co-word analysis by indicating the limitations of some existing methods of handling qualitative data, we shall provide internal validation of the method by considering the stability of the graphs as thresholds are varied and different styles of key-word indexing are adopted. It will be suggested that the method is sufficiently developed to classify articles in a reliable way.

CHOICE OF METHODOLOGY

As was shown in Chapter 7, the matrix of the co-occurrence of words within articles is used as a basis for calculation. Data is initially presented in the form of a rectangular table whose rows represent scientific articles (objects) and whose columns stand for a list of key-words taken from each of these articles (characteristics). By calculating the squared matrix of distances (or similarities) between characteristics, i.e. between key-words, techniques of multidimensional analysis can be applied to this sort of table. Two types of methods are available: correspondence analysis and multidimensional scaling. The limitations of both methods will be briefly discussed.

Correspondence analysis (see Hill, 1974; Benzecri *et al.*, 1973) measures the distance between objects or characteristics according to the principle of distributional equivalence: the distance between two objects or characteristics is a function of the similarity of their profiles in the table. The goal of correspondence analysis is to extract a set of dimensions of decreasing importance in the same way as principal component analysis does for quantitative characteristics. Usually, the representation of objects or characteristics is limited to the space created by the first two dimensions. The relative importance of these first two dimensions in relation to all the others is an indication of the adequacy of this representation.

To apply this method, the file was limited to ninety-four key-words whose frequency was greater or equal to three. Table 11.1 shows that the two first dimensions merely 'explain' 11.2 per cent of the total distances between key-words. Further dimensions have comparable, although decreasing, weights. Altogether, seventeen dimensions are necessary in order to explain half of all the distances among key-words. It is therefore futile to look for a meaning in the first two dimensions.

The explanation of the failure of correspondence analysis shows that it is not a matter of a particular file having special properties. Key-word coded scientific articles never have the usual features of characteristics attributed to objects.

A list of characteristics is generally applicable to all the objects whereas the set of key-words of an article can be of a variable length. This indicates that there is an inherent difference between key-words and characteristics. If the table of initial data is made up of 172 rows and 144 columns, many of the cells will be empty. This does not mean that the indexation was done wrongly. It is due to the difference between the logic of key-word assignment and the logic of the attribution of characteristics in non-documentary databases. In our case, the attributes are used both to retrieve the objects and to qualify them. Key-words cannot be treated as characteristics if the associations between these characteristics must be the combinations of a small number of independent dimensions.

TABLE 11.1 *Correspondence analysis: dimensions that account for 50% of variance*

Dimension	*% explained*	*% total*
1	6.3	6.3
2	4.9	11.2
3	4.1	15.3
4	3.6	18.9
5	3.1	22.1
6	3.1	25.2
7	2.9	28.0
8	2.9	30.1
9	2.5	33.5
10	2.4	35.9
11	2.4	38.2
12	2.3	40.5
13	2.1	42.6
14	2.0	44.6
15	1.9	46.5
16	1.9	48.5
17	1.9	50.4

As the preceding chapters have shown, key-words constitute a language which articulates articles in relation to each other. In this way chains of words appear which are products of translations: they characterise the networks of problematisation that are shaped by translation. Correspondence analysis does not relate to this logic. By contrast from the indexer's point of view, the key-words are obviously heterogeneous: certain ones are very general whereas others, on the contrary, are very precise.

Multidimensional scaling[1] suffers from the same sort of difficulties. The aim of this method is to identify a configuration of words directly such that the calculated distances between the words conform as closely as possible with the geometric distances. This is done within a space which is set at two or three dimensions beforehand. The distances that are calculated between pairs of characteristics are given by one of the indices of association defined in Chapter 7. A coefficient labelled 'stress' measures the residual difference between the geometric distances and the calculated distances.

The possibilities of multidimensional scaling were explored at an early stage of co-word analysis, using a file on basic research and technological innovation in the field of macromolecular chemistry (polymers). It is possible to find some global properties of the field, like the oppositions between physical approaches and chemical approaches, and the distinction between basic and applied research (Callon, Courtial, Turner, 1979). But the results do not allow any more detailed analysis, because the stress is far from negligible (0.30). In fact, the stress would have been much larger if the key-words had not been regrouped into fifty comprehensive categories. In general, multidimensional scaling can produce accurate maps only in there are no more than ten to twenty characteristics. Thus, the applicability of this type of analysis is seriously limited.[2]

The failure of multidimensional scaling seems to be due, as in the case of correspondence analysis, to the fact that the existence of the translation operation is discounted. There may not be a pattern which is common to several actors. The combinations of key-words within articles is determined by the particular configuration of specific interests. These interests are the expression of local logics even if they are occasionally encompassed by more general interests. These logics frequently differ from one another and are related only through the operation of translation. Although one word or group of words is associated with others, the links which unite them are in perpetual movement and the associations between them evolve as a function of the balance of forces. The micro-logical processes may be independent of one another and if they are situated at the opposite ends of a chain they may be completely unaware of each other's existence (cf. Chapter 10).

As a consequence, it is necessary to use more flexible methods like co-word analysis. The data has a variable degree of complexity and it possesses a natural structure in the form of chains. Because of this, the problem of restricting this data to a limited space has to be avoided. Co-word analysis recognises these features by calculating only the values of links between words. These links may be represented in the form of a graph (Gondran and Minoux, 1979). The points of the graph correspond to words; the lines correspond to links that occur more than a certain number of times, or otherwise surpass a certain threshold value.

As was indicated in Chapter 7, two coefficients of the association between words have been adopted for regular use. The inclusion coefficient accounts for the associations between words that are specific on the one hand and those that are general on the other. In this way it is possible to reveal the polarisation of research themes. The proximity coefficient highlights words that occur infrequently. Of course, the two graphs have certain links in common. This will become visible when the thresholds are lowered.

Other coefficients have also proved to be useful for specific purposes.

The Jaccard index which is often used in documentary techniques, expresses the degree of intersection between two sets as:

$$J_{ij} = (c_{ij})/(c_i + c_j - c_{ij}).$$

This coefficient measures the relative degree of overlap between 'semantic areas' of words within a given database. The Jaccard index cannot handle associations between low-frequency and high-frequency words very well, because it will have low values even in cases where the low-frequency word always appears together with the high-frequency word. Therefore, it will only be used to explore overlap between medium-frequency words.

There is another coefficient which, like the proximity coefficient, compares the observed frequency (c_{ij}/N) of a pair of words with the expected frequency of that pair if the words were independent ($c_i/N \times c_j/N$). Compared with the proximity coefficient, it has the advantage of being symmetrical and normalised. This 'statistical coefficient' is calculated in the following fashion:

$$(S_{ij}) = 1/s \times (c_{ij} - c_i c_j/N)$$

where s is the standard deviation of the hypergeometrical distribution (Rip and Courtial, 1984). This coefficient is not used on a regular basis because the strength of association is not an important variable in our graphs: all that is required is that it exceeds a threshold. In addition, calculating this coefficient is costly in computer time, while the extra information is not essential for interpretation.

The flexibility of co-word analysis is apparent from the possibilities of zooming in on certain areas. After starting with a high threshold for coefficients in order to obtain readable graphs, one may wish to see more concerning a specific word of medium frequency. A regular option of the programme of co-word analysis is the production of a sub-graph around a word. This sub-graph contains only those links which lead *from* peripheral words *to* the word in question, and those which lead *from* the latter *to* central words. In this way the sub-graph depicts the total context of such a word very precisely, and is called a 'zoom'. Examples of zooms are given in Chapter 10 for biotechnology patents.

EXTRACTION AND STABILITY OF SKELETON GRAPHS

The inclusion coefficient works on the assumption that the lines of the graph have a certain orientation because the graph is asymmetrical:

words which occur less frequently are included within words which occur more frequently. To facilitate the reading of the graph, the lines are oriented from the word with the lowest frequency to that of the highest.

With the lines oriented in this manner, a topological order can be imposed upon the graph. Each point is assigned a number which is called its rank. If there are only lines oriented away from a particular word, i.e. it does not include any other word, then the latter is attributed the rank of 1. All words that may be reached by oriented lines from rank 1 words are attributed a rank of 2. Ranks are attributed to the words in this way all along the path that leads to the central poles (words reached by lines oriented from rank 2 words are of rank 3, etc.). When a word may be reached by means of several lines which are oriented from words of different lower ranks, the highest of the ranks implied is attributed to this word.

Figure 11.1 shows part of the graph for a subset of the dietary fibre file (1973–6) as an example with words of rank 1 to the right, and a pole of rank 5 to the left. To grasp the basic structure of the map, one should distinguish between the principal trees and additional sequences of precursor/successor words. The link between 'diverticule' and 'colon' in Figure 11.1, for instance, is additional; it may even be the result of the co-

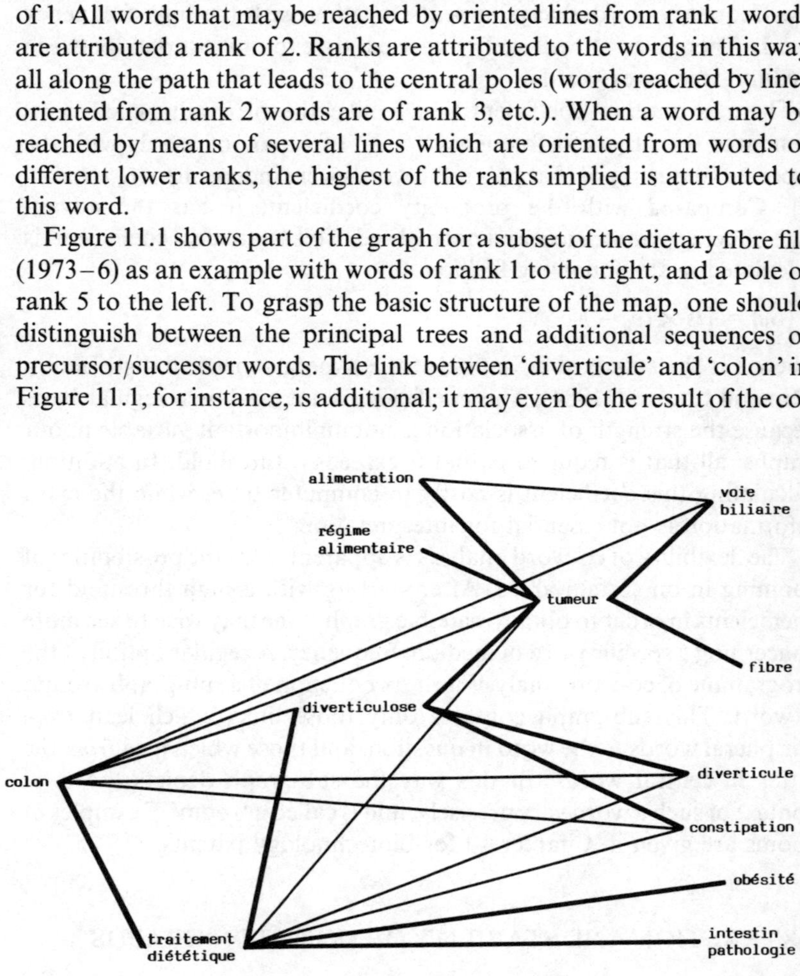

FIGURE 11.1 Detailed graph (all lines) and skeleton graph (thick lines) of part of the dietary fibre graph

occurrences already existing between 'diverticule' and 'diverticulose' and the latter with 'colon'. For the same reason, the direct link between 'colon' and 'diverticulose' is considered less important than the sequence passing through 'traitement diététique'.

Thus, it is possible to simplify the graph without losing much information, by removing all direct links between words that are also connected through intermediaries. The resulting 'skeleton graph' (the heavy lines in Figure 11.1) only contains the principal trees of the graph. The logic of association in sequences of words, going from the specific to the more general, is conserved: all words of inferior rank continue to have access to the words of superior rank to which they were linked in the original, detailed graph (and vice versa). When interpreting skeleton graphs, caution is only required with regard to absence of direct links between words: they may well have been visible on the detailed graph. But this is a small price to pay for being able to highlight the basic structure of the graph.

Skeleton graphs can be used to study the influence of the threshold that is chosen. When the threshold for the inclusion coefficient is lowered, more links are allowed and the number of words on the map will increase. However, as Table 11.2 and Figure 11.2 show, the number of words on the map rapidly reaches a stable value. In the case of the dietary fibre file, the stable value of 75 words can be shown to be reached when the threshold is set at 0.332.

TABLE 11.2 *Number of key-words and links constituting the skeleton when threshold of I is lowered*

Threshold of I	*Number of words*	*Number of links*
0.400	53	54
0.350	73	80
0.300	75	90
0.250	75	96
0.200	75	94

Although this is an empirical finding, there are arguments to consider it a general feature of the Leximappes. For words of low frequency, the threshold operation itself is responsible for the stability of the number of words. Only words with a minimal occurrence of three are taken into account (as explained in the first section of the chapter), and in addition, co-occurrences less than three are neglected to reduce spurious links.

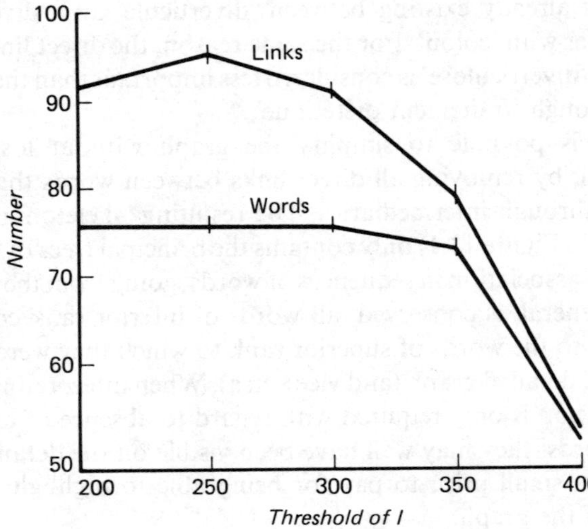

FIGURE 11.2 Stability of the skeleton when threshold of *I* is lowered

Because of the minimum level of three for co-occurrences, low-frequency words will either have rather high inclusion coefficients, or not appear at all. At an occurrence of nine, the inclusion coefficient will still be at least 0.333. Thus, reducing the threshold for the inclusion coefficient below this value will increase the number of low-frequency words only marginally.

For words of high and intermediate frequencies, all values of inclusion coefficients are possible. Clearly, if there is a stationary region when the threshold for the inclusion coefficient is lowered, there must be a structure available in the file where there is a set of key-words with reasonably high inclusion linkages, while other key-words are not linked at all (or negligibly so). It is this structure that is captured by the skeleton graph.

As can be seen in Figure 11.2, the number of links does not stabilise as the threshold for the inclusion coefficient is lowered. It may even decrease because of the way the principal trees making up the skeleton graph have been defined. One new link between high- or intermediate-frequency words may entail the elimination of more than one link between lower- and higher-frequency words (Figure 11.3 provides an example).

The important finding is that the number of links has a ceiling value,

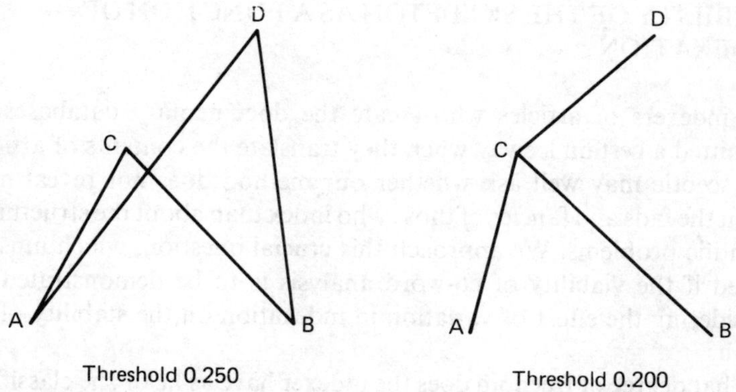

FIGURE 11.3 How an additional link (CD) decreases the total number of links in the skeleton graph

so that the lower thresholds do not increasingly complicate the picture. From the argument about the construction of principal trees, it can be seen that the links between low-frequency words and between some intermediate-frequency words stabilise quickly. The part of the graph showing the structure of the lower ranks is independent of the threshold that is used. The associations between general poles (and from them downwards jumping the ranks) are the least stable. The interpretation of links between frequently occurring key-words has therefore to proceed cautiously: some linkages may result from the increased probability of overlap. In such a case, the statistical index S_{ij} introduced in the first section might be used to check for statistical significance.

The use of skeleton graphs to study structures of files of scientific texts and their evolution over time has been shown to capture a number of essential elements of the logic of chains of key-words. No inconsistencies appear when parameters like threshold values are modified. Therefore, we may use the skeleton graphs to explore other influences, for example that of variation in indexing the texts. Before doing so in the next section, we note an additional feature of the graphs.

Since words are plotted on the graphs by rank instead of frequency, a high-frequency word can now appear at the periphery of the graph: this means that there are few intermediaries between this word and specific descriptors. On the other hand, words of high frequency and high rank are to be found at the top of chains of intermediate words. These general poles represent the best articulated networks of problematisation of the file of texts.

STABILITY OF THE SKELETON AS A FUNCTION OF INDEXATION

The indexers of articles who create the documentary databases are permitted a certain leeway when they translate the contents of articles. The sceptic may well ask whether our method does not reveal more about the fads and fancies of those who index than about the structure of scientific problems. We approach this crucial question, which must be solved if the viability of co-word analysis is to be demonstrated, by considering the effect of variation in indexation on the stability of the graph.

What degree of freedom does the indexer have as he or she classifies a given publication in terms of a set of key-words? In addition to personal idiosyncrasies which can be assumed to average out, variation is caused by at least two structural factors. The first concerns the fact that a given paper always relates to a wide range of other types of texts. In other words, it is situated at a junction within the multiplicity of sub-networks which correspond to multiple networks of problematisation. Therefore it is amenable to multiple translations through key-words. Unsurprisingly, trained indexers therefore vary in their judgements about the most significant context of any given article.

The key-words used to index an article tend to favour certain key issues rather than others. Thus, in the case of diet, some articles are written by doctors and published in medical journals. These relate to a tradition of treatment for diseases of the colon which relies upon 'dietary treatment' or 'enriched diet'. Other articles are written by epidemiologists. These compare African with European populations, and refer to 'diet' but also to 'nutritional deficiency'. Yet others are written by biochemists working on rats. These cite 'nutrition', 'lipid metabolism', and 'diet'. Thus dietary-based treatment for diseases of the colon may, especially if it is drug-based, be indexed under 'dietary treatment' by a physician and 'chemotherapy' by a pharmacologist.

The second structural factor which favours disparity in indexing arises from the fact that within given sub-networks there is variation in the vocabulary. Thus the terms 'roughage', 'vegetable fibre', or even 'cellulose' may be preferred to the notion of 'dietary fibre'. Similarly, the words 'diet', 'nutrition', and 'enriched diet' appear to have similar meanings, even though choice between them depends both on slight differences between articles and on the way in which experimental work is described. The terms 'cholesterol' and 'cholesterolaemia' likewise appear to be similar. Further, when several terms refer to similar issues, indexers usually include them all despite the fact that this may involve redundancy.

For the user of the documentation service this implies that he will retrieve the information he needs whatever key-word he is using. For the co-word analyst, this practice entails a proliferation of key-words located at similar positions in the network, and fragmentation of that part of the graph, especially since such redundancy indexing is not always done systematically.

An additional issue is that indexers are free to include so-called 'macro-terms', global key-words intended to facilitate interrogation of the database. Thus, in addition to the word 'rodent', indexers may add the term 'mammal'. This practice makes it possible for someone concerned with mammals in general to retrieve all relevant articles without listing individual mammals. Again, they might add the word 'lipid' to 'cholesterol' so that articles about the latter would not be overlooked by someone concerned with lipid metabolism. In the case of manual indexation the addition of such 'macro-terms' is not systematic. It may vary between indexers as these differ in their judgements about the appropriate general context within which an article should be located. In practice, however, although such variability does exist, it turns out to be slight.

Macro-terms appear in the PASCAL database of the CNRS in more than 80 per cent of the cases in which a specific descriptor would be sufficient. The internal coherence of indexation is sometimes thought to be so important that it merits introduction of a special coefficient. After indexation of single articles by several indexers, the coefficient is calculated by dividing the number of identical key-words by the total number of words used. However, it is, in fact, far from clear that this coefficient is capable of measuring the quality of indexing. This is because it can be low (i.e. there may be major differences between indexers) yet the 'core' of key-words used can also be identical. The addition of different general poles by different indexers may thus affect the basic meaning of indexing very little, but rather be seen as adding to its quality.

The variability of indexation may influence the quality of the co-word analysis by introducing or deleting linkages in a non-reproducible way. How can one study the manner in which this variability in the indexation affects the skeleton? One way to answer this question is to devise a reproducible indexation, and compare its maps with what has been found in practice. In the reproducible indexation, one word has always been chosen where several synonyms exist, and the same macro-terms

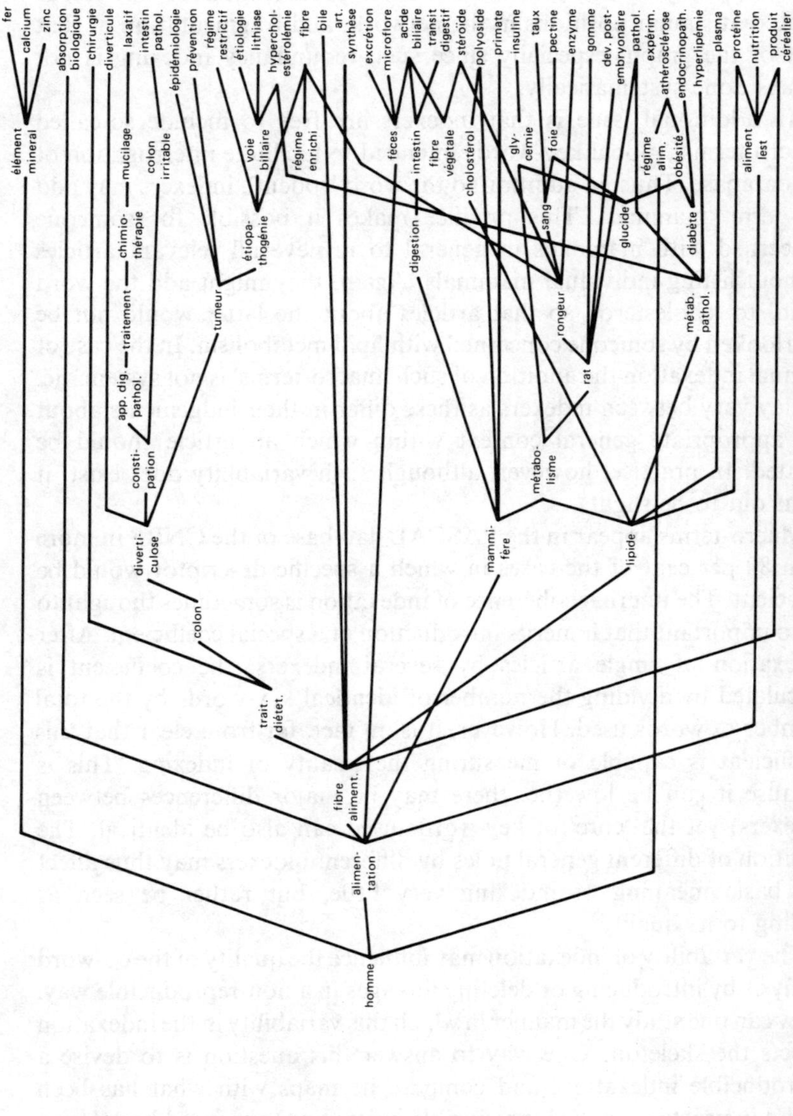

FIGURE 11.4 Skeleton graph of the dietary fibre file, 1973–8

will always accompany the same specific descriptors.

The effect of reproducible indexation can be simulated by reducing the opportunities for indexer variability. In order to create a simulated indexation, the inclusion map for the dietary fibre file was drawn (Figure 11.4) and the general poles, mediator words and peripheral issues were identified. Variation was reduced in two ways.

First, a single word was substituted for a set of similar key-words as they appeared in each article. The following key-words were treated as similar for the purposes of the map:

rat, rodent
cholesterol, hypercholesterolaemia, cholesterolaemia, lipemia, steroid
digestive system, digestive system pathology, digestion, digestive tract, digestive transit, faeces, colon, irritated colon, haemorrhoids, rectum, intestine, intestine pathology
food, nutritional deficiency, nutrition habits, nutrition, diet, enriched diet, restricted diet, dietary treatment
dietary fibre, roughage, cellulose, fibres, vegetable fibres, gum, mucilage, pectin cereal
diverticulosis, diverticule
bile, lithiase, biliary tract, biliary acids
etiopathogeny, etiology
plasma, blood, serum
circulation system pathology, arteriosclerosis
child, man

Second, macro-terms were systematically inserted where these were present in at least 80 per cent of the cases. The following macro-terms replaced narrower descriptors (i.e. those in brackets): lipid (cholesterol); mammal (rat, rodent); treatment (chemotherapy).

The articles re-indexed in this way were the basis for a new map (Figure 11.5). When this last map is compared with that derived from the original file, the conclusion must be that the two approaches depict the same basic structure. There are two main components in each case. One is concerned with the treatment of diseases of the human digestive organs. The other centres around the study of lipid metabolism in mammals. These two components are both related to nutritional habits and dietary fibre. Within each component there is the same network of intermediate poles, e.g. 'diverticulosis', 'constipation', and 'tumour' on the one hand and 'metabolism', 'pathology', and 'glucide' on the other.

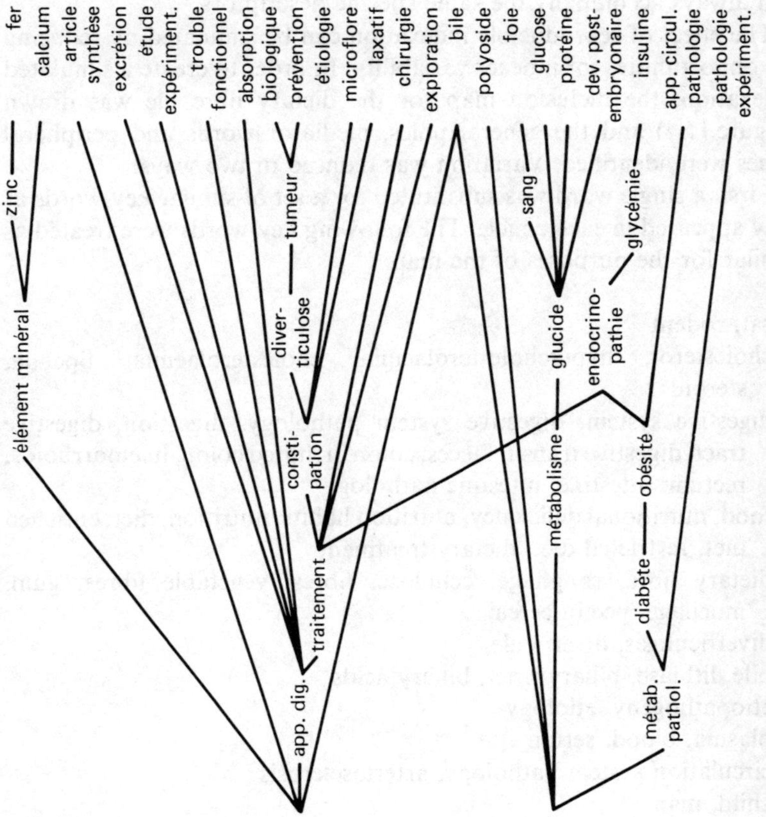

FIGURE 11.5 Skeleton graph of dietary fibre file, re-indexed to reduce variety of descriptors

Naturally many links between macro-terms and the narrower descriptors have disappeared in the simplified map.

Although shown for the case of dietary fibre only, one may infer from this example that the structure of the inclusion map is little affected by indexer variability as this is expressed in the deletion or insertion of macro-terms or the substitution of single words for sets of near synonyms. In particular, mediator words, and small sections of the graph remain stable (see Courtial and Callon, 1985). Maps that make use of a stabilised and simplified vocabulary are relatively simple and hence relatively easy to interpret. The fact that such 'simplified maps' drawn on the basis of a 'corrected' vocabulary differ little from the original inclusion maps points to the validity and robustness of the method of co-word analysis.

If the aim of analysis is the identification of research issues and their interrelations it is often preferable to retain the various layers of meaning and redundancies of maps that are based on initial 'uncorrected' indexation. For instance, links may appear initially between 'dietary fibre' and 'tumour of the colon', while 'vegetable fibre' relates more directly to research concerned with digestion. The progressive development of a connection – that is, the gradual amalgamation of the concepts 'dietary fibre' and 'vegetable fibre' – becomes the end point of a process in which the two previously unrelated research areas are brought together.

INDICATORS OF STRUCTURE

There are many indicators available in graph analysis to describe the structure of the graphs that are produced in co-word analysis, for instance degrees of connectedness, the occurrence of 'stars', i.e. points with many links (compare the examples in Chapter 8), etc. It is often necessary, however, to devise specific indicators of structure. One reason is that the maps reflect the patterns of the networks of problematisation available in a scientific field, which may well be specific to that field. Also, one may have different interests, ranging from comparing networks of problematisation across fields to analysing particular shifts in a field and identifying actors that carry these shifts.

In this section, we shall discuss some examples of indicators of structure that are peculiar to maps of scientific fields, in the sense that they are based on the importance of chains of key-words in the maps (compare with the first section of this chapter). As a first step, one may distinguish elementary patterns of key-word networks (Figure 11.6). Linear chains, i.e. sequences where each key-word has only one precursor and one successor at most, are possible, but do not occur very frequently. More often, the chain is branched, as in the monopolar network of Figure 11.6(b). Since branching provides options for interaction between problematisations, key-words having more than one precursor and/or successor can be distinguished from those that have not, and their ratio may be an indicator of the potential for change in the network.

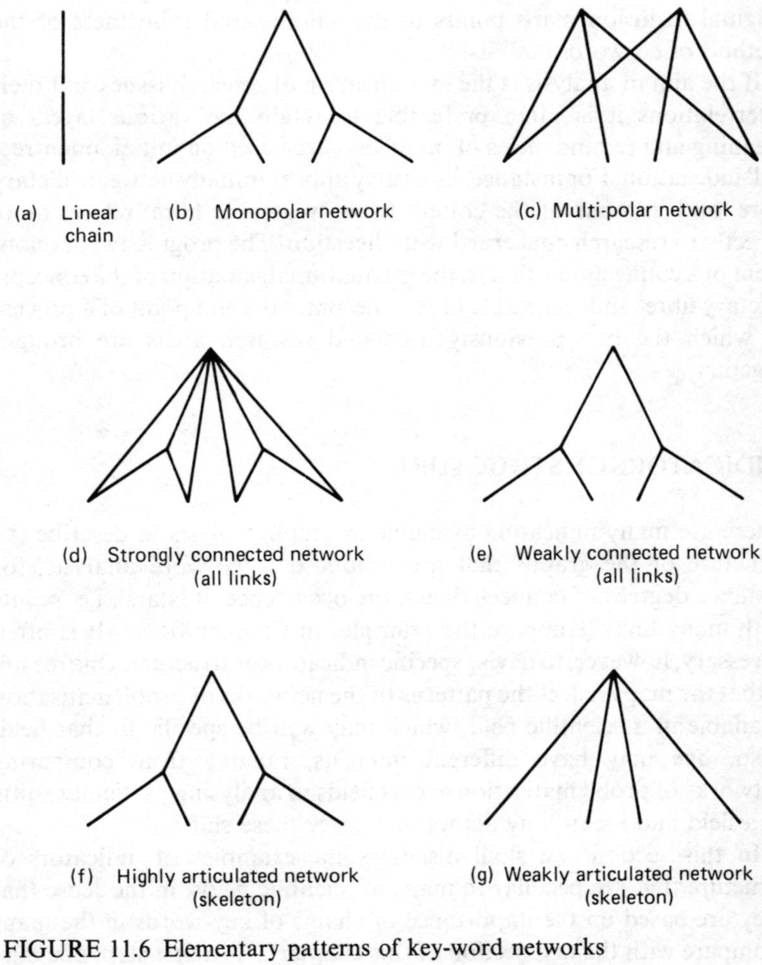

FIGURE 11.6 Elementary patterns of key-word networks

When branching occurs, there still is a difference between weakly connected networks (Figure 11.6(e)) and strongly connected networks (Figure 11.6(d)). A useful measure of the difference can be constructed by comparing both with the skeleton graph that can be extracted from them. In these examples, the skeletons coincide, and are equal to the highly articulated skeleton graph in Figure 11.6(f). The number of links that are removed when going from the original map to the skeleton is a measure of the distance between the two, and thus of the interconnectedness of the original map; for the examples in Figures 11.6(d) and (e), the numbers are four and zero, respectively. The importance of this measure is that it reflects the connectedness in the scientific field. When specific descriptors are linked by a mediator word, as in Figure 11.6(e), they need not belong to the same problem area. But if both are also linked to the same higher-level pole, one may assume that there is more integration. Thus, the distance between original map and skeleton graph is an indicator of (lack of) fragmentation of the scientific field.

Key-words can be classified according to their position in different types of chains. Words that lead to several central poles are particularly important; we have called these passageways. They correspond to articles which introduce new links that cannot be reduced to the trees of each pole taken separately. The number of passageways for each pair of poles can be an indication of restructuration of multi-polarisation in the field, as, for instance, that which occurs with the development of new research fronts which relate to two different disciplines. They can also correspond to new problems that have repercussions for several different fields.

SORTING BY MEDIATOR WORDS

Inclusion maps have been shown to distinguish between specific key-words of low rank, mediator key-words on which links from other key-words converge, and central poles, general key-words that terminate numerous chains of successive links. In this way, it becomes possible to determine the specificity and generality of key-words more or less independently of the indexation procedures, and to identify mediator words that can be used to classify the scientific texts in a useful way.

The problem is important. When articles written in different countries, laboratories and journals are compared on the basis of their contents, only the key-words of the indexation are available. Since the frequency of key-words is quite variable, they do not provide a very good basis for easy comparison. If general key-words are used, only vague results are obtained. If specific key-words are used, they are found to be absent from most of the articles. What is required is a selection of existing key-words at the same level of generality (with respect to the scientific field that is being studied) and which do not overlap too much in content.

The inclusion maps provide a systematic answer. They show that there are a number of 'field-wide' words, words of a very general nature (often included as macro-terms by the indexers) or words that designate the nature of the field that is being studied. In the case of the dietary fibre file, 'dietary fibre', 'nutrition' and 'man' fall into the category of general words, while 'cholesterol' and 'lipid' have to be considered 'field-wide' words, because their content is similar in this particular field (they are situated in the same chain). Therefore a list of 'field-wide' words must be decided by an expert, with the aim of deleting words which do not contribute to significant differences among articles.

After eliminating the 'field-wide' words, it is easy to select a set of mediator words that do not overlap in content. (If necessary, words that are semantically close to each other can be merged into one.) The scientific texts can then be sorted by these mediator words that occur in their indexation. A computer programme MULTIPOL is available for sorting; in the case of dietary fibre, the results are shown in Table 11.3.

TABLE 11.3 *Sorting articles by mediator words*

Group	*Mediator words*	*Number of articles*
1	Métabolisme	37
2	Métabolisme–Digestion	3
3	Métabolisme–Métabolisme pathologie	2
4	Constipation	19
5	Constipation–Métabolisme pathologie	1
6	Constipation–Métabolisme pathologie Tumeur	1
7	Constipation–Tumeur	5
8	Digestion	20
9	Métabolisme pathologie	16
10	Métabolisme pathologie–Tumeur	1
11	Tumeur	12

Of course, only articles indexed by at least one of these mediator words are taken into account. In the case of dietary fibre, 117 articles out of the file of 172 could be classified this way. The remaining articles can be analysed by other means, or by re-coding them with the help of the notion of potential poles.[3]

In the case of dietary fibre, most articles fall under one of these mediator words only. A by-product of the MULTIPOL analysis is the identification of articles with combinations of mediator words, sometimes rare (original?) ones that can be subjected to qualitative analysis. An article combining three mediator words for the dietary fibre field is indicated in Table 11.4.[4]

TABLE 11.4 *Example of an article exhibiting a rare combination of mediator words*

Title	*Key-words*
L'Hypothèse de Burkitt-Trowell concernant la déficience de 'matières cellulosiques diététiques' dans la ration quotidienne	Endocrinopathie Diabète–obésité Appareil digestif pathologie Interaction–déficit Fibre alimentaire Cellulose Constipation Diverticulose Colon-tumeur Etiopathogénie Homme Régime alimentaire Métabolisme pathologie Fibre végétale

In other cases, combinations may be more frequent, and the mediator words are, in a sense, less independent. However, it will always be possible to create groups of articles relating to single words or combinations, and do further analysis in terms of these groups (provided they are not too small). For instance, one could trace the distribution of countries of origin of the articles within and across these groups, or identify laboratories and scientific journals that play important roles.

As an example, Table 11.5 shows the activity profiles of the four main countries involved in dietary fibre research. Some striking differences like the inversion of the activity profile of France on the metabolism and constipation poles, compared with the other countries, may be related to the overall structure of the map, with two main components: one concerned with diseases of the human digestive organs (and clearly

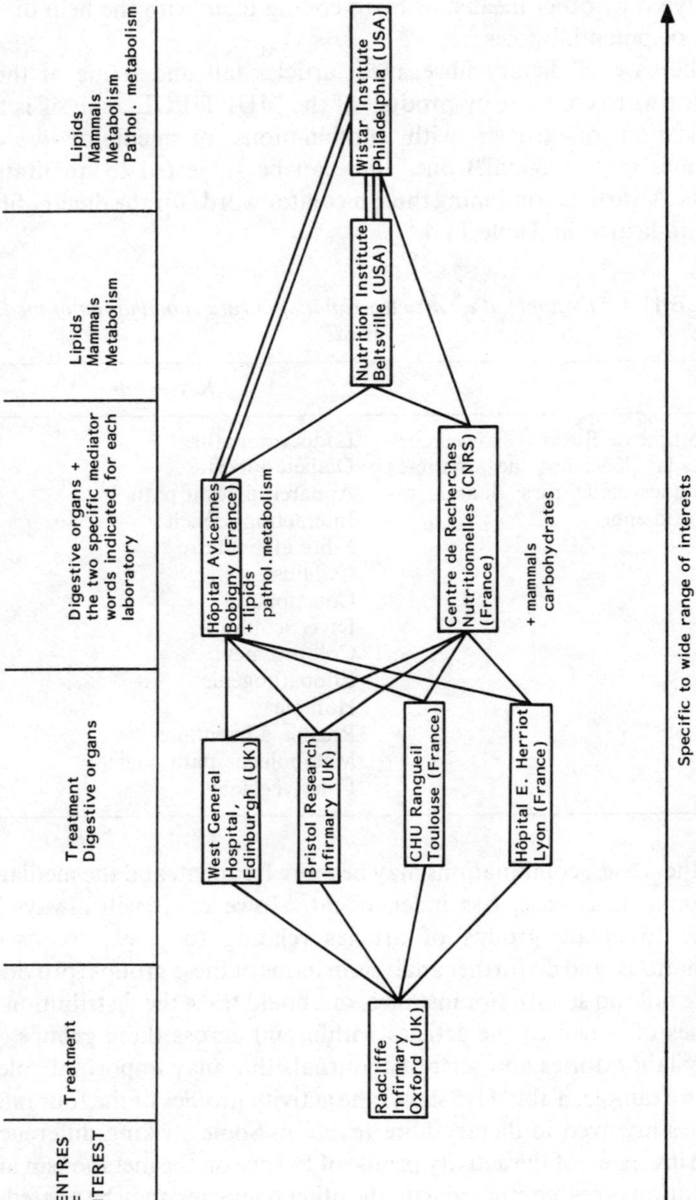

FIGURE 11.7 Laboratories in the dietary fibre field

TABLE 11.5 *Activity profiles of countries, measured by numbers of articles falling under different mediator words (in percentages of number of classified articles per country)*

	Métabo-lisme	*Consti-pation*	*Diges-tion*	*Métab. Patho.*	*Tumeur*	*Total (%)*
USA	45	6	19	15	15	100
UK	32	14	18	14	22	100
France	3	47	17	14	19	100
Germany	43	13	31	13	0	100

important in France); the other focused on lipid metabolism in animals (see the third section on skeleton stability and Figures 11.4 and 11.5).

Further possibilities for analysis may arise as, for instance, in the case of dietary fibre research when the mediator words are relatively independent and can be interpreted as research areas. The research interests of the main laboratories active in the field can then be described by specifying the mediator words that predominate in their publications. Also, overlapping research interests can be plotted as linkages between laboratories. Figure 11.7 gives the results for dietary fibre research. The two main components of the field are visible again. Further articulation is possible by looking in more detail at the articles published by the laboratories.

The examples from the dietary fibre file shows the potential of classifying and sorting articles on the basis of those key-words that have been selected as mediator words. Although the inclusion maps from which these words are derived have to work with the raw material provided by documentation services, it turns out to be possible to improve on it by considering the patterns of links to be more informative than the existence and frequency of occurrence of specific key-words. As we concluded in the third section of this chapter, the structures revealed by co-word analysis are more robust than the variability of the input data might lead one to think at first glance.

NOTES

1. MDSCAL, KYST programmes (Kruskal, Young, Shepard, Torgerson), Bell Laboratories, 600 Mountain Avenue, Murray Hill, New Jersey, 07974, United States.

2. In addition, the validity of one of the presuppositions of co-citation analysis is called into question. This hypothesis claims that whatever the degree of stress a socio-cognitive two-dimensional space exists.
3. When articles are not classed by mediator words, they are indexed by key-words which are subordinated to them on an inclusion map. In this case, these mediator words can be considered as potential poles of the articles, despite the fact that none of them appear explicitly as an indexing term. A computer programme was developed which automatically classes articles with respect to their potential poles.
4. A programme called DISTOR exists which takes into account the mediator words to which article key-words lead, according to the key-word graph, even if the pole is not explicitly mentioned. The programme then ranks all articles according to the degree to which they contain remote poles.

12 Future Developments

MICHEL CALLON, JEAN-PIERRE COURTIAL and WILLIAM TURNER

Compared with other techniques such as co-citation analysis which already has a long history, the co-word analysis technique is only now entering its development phase. More work will be required before it can be used in a routine manner. In the first part of this book we presented the theoretical foundation of the method. Both the translation and the actor-network theories will be further elaborated elsewhere in order to show their full implications for the social sciences. But these theoretical developments will probably have only a limited impact upon the co-word analysis technique. Improvements in the method will essentially depend upon how it is used and the reactions which are encountered.

Since initial testing of the method in a policy context has indicated that there are a number of problems that require resolution (Healey and Rothman, 1984) we have undertaken a programme to improve both the indexing techniques and the methods used to interpret the maps.

TOWARDS AUTOMATIC INDEXING

The use of the co-word analysis technique is interesting because at the present time, with the exception of the Institute for Scientific Information in the United States which uses citation indexing for document retrieval, all other database producers use some form of key-word indexing. An estimated 4 155 000 articles and documents will be treated in 1985 by the American indexing services alone. This wealth of information, if properly exploited, can provide systematic information about changing centres of interest in scientific and technological research and, at the same time, serve to identify who is working on what in specific problem areas.

Our experience with the co-word analysis technique has shown that the exploitation of manually indexed databases raises three major problems. The first is that certain types of literature are not generally indexed. This is particularly true of patents and the 'grey literature' which usually lies outside the publication circuit and includes, amongst other things, congress proceedings, doctoral dissertations, reports, science policy committee meeting minutes, laboratory records, notes and correspondence.

The second problem is the time which elapses between the writing of a document and the moment when it is indexed and entered into a database. This delay can be quite substantial, depending upon the journal in which an article is published, and referencing and indexing practices.

The final problem is what has been termed the 'indexer effect'. It concerns the way in which key-words are related to documents in the databases analysed. Several factors are involved: the content of the documents analysed; the rules adopted by database producers in order to structure document retrieval strategies; and the intellectual background of the indexer who is asked to identify the problems raised in an article given his or her general knowledge of a research field.

It is important to find answers to these three problems. Patents must be analysed in order to develop tools for technological decision-making and to explore the links between fundamental and applied research. An analysis of the 'grey literature' offers quicker and perhaps more direct access to what is happening on the research front, because science policy decisions, laboratory notes and records, letters and congress communications are the raw materials used to produce publications. Time lags reduce the possibility of developing 'alert systems' for identifying emerging research themes. The indexer effect concerns the quality and reliability of the results obtained.

The evolution of what is now called the information industry has led us to believe that the solution to these three problems lies in the development of computer-assisted indexing aids.

The Changing Information Industry

Perhaps the most noteworthy change in the information industry is the fact that all the elements of the chain from word processors to the electronic delivery of documents are now interconnected. Some writers speak already of the 'paperless society' (Lancaster, 1982). The project under way at the Patent and Tradesmark Office (PTO) of the United States illustrates what is involved (Commissioner of Patents and Trademarks, 1983).

The PTO's task is to issue patents and trademarks for the United States. This activity has generated a database which at present contains about 24 000 000 documents. In order to automate its activity, the PTO has taken three decisions. First, most paper files and paper handling operations will be replaced by electronic and other computer-accessible databases. Second, communications with applicants and constituents will be in modes and media chosen by the latter and supported through electronic terminals, telecommunications and other automation technology. Finally, most patent and trademark pre-examination, examination, post-examination, classification, and management control functions will be supported through electronic work stations and automated or computer-assisted operations.

Given the evolution of technology, full-text electronic databases will become increasingly current. Scientific and technical editors such as Springer, Elsevier, John Wiley, Blackwell and Pergamon Press are studying their creation in connection with document delivery systems which will enable them to supply articles upon request rather than in the scientific journals which we now know. The European Commission, in the framework of the ESPRIT (European Strategic Programme for Information Technologies) project, is sponsoring a Community-wide research effort on new ways of disseminating scientific and technical information which could radically change the character of scientific writing.

For example, electronic journals might make it possible to restructure peer review. At present, the time taken to referee papers is a factor in delaying the dissemination of information. In a situation where the members of a scientific community are linked together in a network and, consequently, are in a position to see how others comment upon, criticise or suggest modifications to a text which has been made available to them through the network, refereeing becomes a community affair rather than the concern of a limited number of experts or gatekeepers (Ducasse *et al.*, 1985).

Host services for database producers such as TELESYSTEMES in France or MEAD in the United States have drawn the conclusions which are appropriate in this context of rapid change in techniques and ideas about the dissemination of information. They are developing full-text indexing techniques for information retrieval. The problem with the mass of information which is now starting to circulate electronically is that of locating what is significant for the individual user, given his or her particular activity. This is precisely the task of those host services in cooperation with the various database producers they represent.

The Advantages of a Semi-Automated Full-Text Analysis

The co-word analysis technique is currently being developed in collaboration with TELESYSTEMES in France. As a host service, TELESYSTEMES is responsible for the on-line commercialisation of a wide variety of scientific, patent and grey literature databases. Consequently, we have access to the databases needed to develop scientific, technological and industrial decision-making tools. These databases are updated rapidly. For example, Derwent Publication Limited offers its patent database through TELESYSTEMES. Derwent receives over 11 000 patent documents a week and updates its files weekly.

The full-text information retrieval software used by TELESYSTEMES is known under the name of QUESTEL-PLUS. Single words are identified and located by means of a code as being within a given sentence of a paragraph in a particular section of the document. The spelling of words is not standardised, nor is there any attempt by QUESTEL-PLUS to eliminate synonyms and those words other than nouns which do not designate specific research subjects. We have been studying these problems and techniques are currently being tested which are capable of interactively standardising the spellings of nouns, while eliminating all other word forms.

Three different techniques can be used either separately or in combination with one another. The first is to create a dictionary and then compare the words it contains with the word list automatically generated by the QUESTEL-PLUS programmes. The second is to supply experts with suitable software in order to enable them rapidly to select the nouns which designate what they consider to be significant research subjects in their field. The third is to systematically eliminate words whose frequencies in the data file are below a certain threshold.

When dictionaries and statistical procedures are used together, fully automated indexing systems may be developed. In terms of the co-word analysis programme, this offers the perspective of a completely computerised chain of procedures running from a QESTEL-PLUS treatment of full-text literature to the automatic establishment of inclusion or proximity maps by the LEXIMAPPE programmes. The software permitting rapid expert intervention is necessary as a means of controlling the proper functioning of the chain.

We have tested the chain on a small dietary fibre file. The results obtained are encouraging because it would appear that we are in the process of overcoming two difficulties we have had with manual indexing. Indexing services often use very general terms and multiply synonyms in order to increase the numer of retrieval strategies that can be used to find a given document. The words used do not necessarily appear in the texts themselves. They serve as a means of defining the contents of a document with respect to what the indexing service considers to be different themes of the research under way in the field. On the other hand, specific subjects treated in the documents are often ignored because they are considered too precise for document retrieval purposes. As shown in Chapter 11, these distortions do not affect the general structure of the inclusion maps.

When treating the automatically indexed dietary fibre file, we found in comparison with our manually indexed file that: (a) the general and redundant words which complicate the reading of the maps without adding new information were eliminated; (b) a much larger number of specific peripheral issues were found on the inclusion maps; (c) the structure of the proximity maps was much richer and more detailed.

With the development of this software, we should be able to considerably reduce, if not eliminate, the 'indexer effect' in presenting the results of a co-word analysis. But there is another, more fundamental, reason for moving towards the exploitation of full-text databases. It is to do with changes in the information industry. We will soon be in a situation where full texts will be very quickly available for analysis. The extremely rapid circulation of information which the new information technologies imply make it possible to envisage the development of up-to-date 'alert systems' for decision-making in science and technology. The goal of the QUESTEL-PLUS/LEXIMAPPE research programme is to use the co-word analysis techique in order to produce these systems.

HOW TO IMPROVE MAP INTERPRETATION: THE COLLABORATION WITH THE EXPERTS

As we have underlined on several occasions, the maps obtained by co-word analysis have to be interpreted with caution. In particular, a map considered by itself is generally very difficult to understand. The method of interpretation must necessarily be active and presupposes the comparison of inclusion maps and proximity maps, the use of zooms and, if necessary, an examination of the original documents or, at least, their descriptors. The co-word analyst can carry the work of interpretation to a certain extent, but cooperation is difficult to establish. As Healey and Rothman (1984) have noted: 'A paradox exists in the validation of science policy indicators. If the results of the work are counter intuitive to experts they are considered invalid; if intuitive, they are considered valid but uninteresting – they reveal only that which is already known.'

No matter what the degree of expert scepticism is, one of their natural tendencies is to take the maps at their face value and limit their attention to a close reading of the words and their associations. Often they are disappointed not to find the subjects which they know to be important for their field described in the detail they would expect.

However, as the argument of this book reveals the goal of co-word analysis is not to photograph a field of knowledge but to reveal the strategies by which actors mutually define one another, relate and place their problems in a hierarchy. If one is to grasp the strategic dimensions of a co-word analysis, the maps cannot be considered statically. They must be interpreted dynamically.

We have progressively come to the view that the maps are not the final product of a co-word analysis. A whole series of documents are needed to identify the passageways, to signal and explore the problem areas, to list and explain the 'deviant' articles, or to trace the activity profile of a country, laboratory or firm. These documents take as their starting point the changes which are shown on the maps. Such changes are not to be taken at face value. As we have shown, particularly in the patent study, the disappearance of a central pole must be investigated otherwise hasty and probably false conclusions will be drawn. The investigation calls for proximity maps, zooms, the identification of actors and their strategies given the key-words they associate.

Co-word analysis differs from other methods by the importance it places on the dynamic restructuring of scientific activity over time. In the time to come, we expect to be able to develop a standardised procedure which will render map interpretations more systematic and easier. This will make it easier to collaborate with experts. We want to be able to provide them with hypotheses concerning network transformations and actor strategies; their reactions will supply us with a new starting point for other questions.

Establishing a satisfactory working relationship with the experts requires a good deal of thought. Considerable progress should be made by organising discussions of the development of a field, rather than around maps which are difficult to understand. Their participation in delineating the databases which are to be co-word analysed should also be actively sought, as should their advice on the document retrieval strategy that could be used to constitute the data file. This is an important moment in the co-word analysis programme and every attempt should be made to ensure that differing opinions are expressed. Finally, as a way of stimulating discussion, before the experts are presented with the results of the co-word analyses, they could be asked to identify what they consider to be the strategic sector of their activity and the actors responsible for the dynamic of their field.

FINAL REMARKS

In Part III of this book, we have been basically concerned with decision aids for scientific and technological policy-making. However, sociologists and political scientists who study the dynamics of science and technology might well eventually be able to use this co-word analysis technique as a tool for their analyses. This is because its theoretical rationale is largely inspired by recent studies of the social construction of scientific and technical knowledge.

The method should also be of use in the documentation field. It can serve as a means of classifying documents in terms of their evolving centres of interest and should, from this point of view, be of use both for retrospective retrieval and the constitution and updating of thesauruses.

Part IV
Conclusion

13 Putting Texts in Their Place

MICHEL CALLON, JOHN LAW and ARIE RIP

In this book we have mounted an argument about science and technology and offered a method for its study. The implications of the argument, which were developed in detail in early chapters, are reviewed for science studies and science policy studies in the first section of this chapter. The method, that of co-word analysis, which we see as being the pursuit of the qualitative argument by quantitative methodological means, is assessed in the second section.

However, though most of the present book is concerned with the detailed analysis of laboratory activities, scientific texts, and ways of creating maps of signal words that reflect the structure and evolution of scientific domains, the theory of the actor-network is of general applicability. In the final section of this chapter we touch upon these broader concerns by considering what the approach implies for the status of social science theorising, the distinction between the 'macro' and the 'micro' in social theory, and the analysis of power and social control. In previous chapters these general themes have been implicit rather than explicit since we did not want to burden our presentation with extensive critical discussion of the literature. In the same spirit we here articulate what is distinctive about our approach and not what we take to be mistaken about the approaches of others.

SCIENCE, POLITICS AND TEXTS

Let us start, then, by returning to a theme broached in Chapter 1. In contemporary, industrialised societies, the laboratory (and its equivalents) are strategic loci for social change. The Menlo Parks and Schenectadys of our world have to be treated in the same way as the Fountains Abbeys or the Doge's Palaces of yesterday. Yet, though everyone in Renaissance Italy knew that politics was pursued by Princes in Palaces, it seems to be less obvious that politics, in the sense of work aimed at or contributing to social transformation, is now undertaken in laboratories. There is, of course, an awareness of the fact that there is interaction between science and politics. Indeed, a discipline of science policy studies has come into being to describe and analyse such interactions. But the relationship between the two is a deeper one and requires an approach which goes further than the concerns about big science and the setting of priorities that dominate most science policy studies.

It has been our contention that science should be viewed as politics pursued by other means (Latour, 1983). If politics is about power then, for science, power struggles concern not only negotiations for external resources and internal battles for social and intellectual domination. They have also to do with exerting force to assemble, or disassemble, apparatus, materials and texts. In other words, they have to do with the actual production of scientific facts. Perhaps it is this that constitutes the defining characteristic of science and which offers an explanation of its power in effecting social transformations.

The particular activities in a laboratory, the fitting together of bits and pieces, exert an influence far beyond their local setting because they are linked through chains of equivalence to the large scale and the general. It is in this way that laboratories are able to bend the itineraries followed by others and enrol distant actors in their worlds. This is how electrons in a fuel cell (Chapter 2) become a political entity. They do not *determine* the outcome of political decision-making, as technocratic ideology would have us believe. Equally, however, decisions are not to be understood as free choices between all the possible options. Some worlds, and some links, are more powerful than others. The entities produced by science – for instance polymers, electrodes, rats being fed carcinogens – cannot be easily set aside. The decision-makers have to accommodate to their presence. The laboratory tends to become an obligatory passageway if power is to be exercised.

Thus a double redefinition of scope is necessary: a political approach to scientific practices has to be complemented by a cognitive approach to science policy-making. Science policy studies, like science policy-making, neglect the content of the sciences at their peril.

How should such a redefined study of science and society proceed?

The answer is that it is necessary to cut through myths about the gulf which separates science and politics. In short, double standards must be avoided. Just as Machiavelli's Prince was not a schizophrenic, authentic within the palace but perfidious outside, so too the scientist must not be seen as a disinterested intellect within the laboratory and a politician outside. Action is continuous, and the fact that parts of it are black-boxed and made unassailable by being called 'disinterested' or 'authentic' is itself a part of that action.

It is necessary to follow Machiavelli's example. The actors must be followed without fear or favour. We must enter the strategic loci and describe the strategies that are pursued within their walls. And, we repeat, if our descriptions suggest that science operates in ways that do not conform to its public image, then so be it. Our aim is not to pass judgement. Rather it is to reveal the forces that are at work.

What has been discovered about the power of science? Within the laboratory there seem to be three main activities: discussion, writing, and what might be called manual handling. Each of these can be seen as a collection of appropriately arranged bits and pieces, the creation of a local world. The important thing about these worlds, however, is that they are designed to act at a distance. Emissaries are sent out to conferences, machines and instruments are transported, and money is transmitted. And, perhaps most important, texts are also dispatched. The laboratory is a locus where force is concentrated and packaged so that it can be sent out to exert influence at a distance.

We are not saying that science and technology can be reduced to texts. This is obviously wrong. Rather, we are suggesting that texts, often in combination with bodies or machines, constitute a crucial method of long distance control in science (see Law, 1985a). There are physical reasons for the importance of texts: they travel well (at least since the advent of the printing press and a reliable postal service) and they do not, on the whole, degrade en route. The way in which their power is exerted depends, however, on the way in which they encapsulate the world that their author wants to build. They juxtapose elements, suggest their appropriate relationships, and they simultaneously make an argument about how the reader should fit into that world. Properly built they thus constitute a formidable class of translation operators.

Because texts, by their very nature, display their structure in order to exert their force, they are also particularly amenable to analysis. There are qualitative methods that highlight the documentary strategies of scientists–politicians. And there is also the possibility of quantitative analysis, if suitable ways of distilling and aggregating the force of words can be found. In the next section we discuss the method that makes this possible. Here we wish to emphasise the heterogeneity of the worlds displayed in scientific texts, a heterogeneity that reflects the way in which polymers, electrodes, social problems, scientific institutions and social networks of scientists are mobilised to build such worlds. In order to describe such heterogeneous worlds and their dynamics in general terms, we introduced the notion of the actor-network. This concept is important in part because it overcomes the macro–micro distinction: actor-networks may either grow, or decline, in size. Indeed, the strategies of scientists, as well as those of other actors, can be characterised as attempts to make relevant actor-networks grow.

This theoretical perspective has implications for science policy. Astute observers of the science policy scene have sometimes noted how little power policy-makers actually have, how they cannot do more than negotiate and modulate a polycentric system of forces. As with politics in general, the balance of forces is in continual flux but the rate of change may be higher because scientific developments form an integral part of those forces. Such observations have led to cynical resignation, or to 'garbage can' theories of decision-making. But the forces involved are not random in form: some entities are more durable than others and their linkages are more likely to be the starting point for effective intervention. The concept of the actor-network is ideally suited to the required policy analysis. Its attendent conceptual apparatus and vocabulary may not be necessary in all cases. Decision-makers and their advisers may be intuitive strategists. But the perspective is necessary to avoid being thrown off course by non-existent boundaries between science and power.

In retracing the struggles between actor-networks, the analyst is not simply reducing opacity. Like Machiavelli when he published *The Prince*, he is also affecting the balance of power. Even within the social studies of science the power struggle can be recognised. The debate about externalism versus internalism is a replay, on another level, of the struggle between the palace and the laboratory. Which is to be counted as the 'cause', as the determining force, and thus be protected from interference in its inner workings?

Dissolving opacity is not just a matter for policy-makers and science analysts. To show that there is no mystery surrounding science is to change our Western culture, or at least to place this possibility upon the agenda. Again, this implies a shift in power. To give a very concrete example, will co-word maps enable lay people to ask intelligent questions about science? In other words, will they be able to interfere?

This particular tool has yet to become sufficiently robust and widely available to become a passageway for new groups. But the general point remains: What happens if analysts provide perspectives, descriptions and tools that dissolve opacity and reveal the continuity between scientific and political action?

The answer to such questions depends upon the robustness of the analytical tools. The conceptual apparatus introduced in this book will be taken up primarily by other analysts of science, and may perhaps gradually seep through into the vocabulary of practitioners. However, the maps of forceful words are, at least in principle, more immediately relevant for broader audiences. It is to an assessment of their robustness that we now turn.

THE CO-WORD METHOD

The method of co-word analysis has been developed to trace the struggles between scientific actors as these become visible in the texts of science. The maps not only present the state of an area at a particular moment, but also make it possible to follow the development and identify some of the dynamics of such struggles. The qualitative problem of distinguishing between the successful and the less successful strategies of translation is solved by quantitative means: the aggregation of the co-occurrences of signal-words across a population of texts and the depiction of significant levels of such co-occurrences by graphical methods.

The nature and limitations of the methodological assumptions and the indices used in the co-word analysis have been discussed at length in earlier chapters, and we shall not consider the technical questions again here. Overall, the method has clear advantages, but also some disadvantages.

Compared with other methods of analysis that focus on texts, co-word analysis is much more flexible, accepting any database that can be expressed as strings of signal-words. Also, since each link can be traced back to specific texts using these signal-words, one may trace the co-word patterns and the corresponding strategies of actors in defining and redefining the field in as much detail as one wishes.

By having recourse to the quantitative, it thus becomes possible to map the degree to which the efforts of actors to build their worlds are met with success. The maps present the pattern of translations that arises from the interaction between the efforts of many actors.

Using the quantitative in pursuit of the qualitative, we are also able to highlight features of scientific fields that have not always been recognised. The heterogeneity of scientific world-building is preserved in the co-word analysis, where experimental findings, research methods, concepts, social problems, material artefacts and locations may appear together on the maps. In addition, a profile of central poles (words that assemble many other words), mediating words and peripheral issues can be used to display the cognitive polarisation in a graphical manner.

Perhaps the most important feature of the method is that it makes possible the discovery of chains of translations – that is to say, mediating words that link poles which might otherwise seem to be independent of one another. In a number of cases such chains of translations have been shown to be the basis for a rearrangement of the cognitive profile of the field. In the past, such processes of scientific development have often been neglected because the focus has been on trajectory models of scientific development in which a paradigm or breakthrough finding is articulated as time passes. Indeed, co-citation clustering often assumes the trajectory model when interpreting the evolution of clusters. Naturally, there will be cases where a trajectory model is applicable, but it is our conviction that many areas of contemporary science fit only with difficulty into this model. These are better analysed as a network of crisscrossing chains of translation and, indeed, even fields displaying a trajectory are equally susceptible to such an analysis.

However, the limitations of the co-word method should also be recognised. The dependence on indexation is one such problem: indexing is an intervention between the text and the co-word analysis, and the validity of the maps will depend, to a certain extent, on the nature of the indexing. Yet since indexers try to capture what it is about a text that is interesting, they partially reproduce the readings that the texts are given within the field itself. Thus, despite the fact that indexing is not entirely reliable, validity is never totally absent.

An acceptance that the maps are interesting by scientists and science policy-makers and that they may mark a contribution to their endeavours, is important but not decisive. This is because rejection may result from the fact that the maps show features of the field that are unknown (and/or threatening) to scientists. If we follow the actors without fear or favour this does not imply that our findings will be immediately recognisable to them. The dynamics of the chains of translations that arise from the interaction between (simplified) actor-worlds are sometimes perceived only with difficulty by any individual actor.

This is not to say that the interpretation of the maps is easy. Care is called for, and it will normally be necessary to use several different maps and zooms to explore the features of a co-occurrence matrix. In addition, it is also important to be sensitive to the artefacts with which the maps are strewn – artefacts which occur not because the maps themselves are unreliable, but because they frequently occur themselves in the scientific literature.

Even when the limitations of the method are borne in mind it provides an important tool for the sociology of science and science policy studies. This is not only because it produces interesting maps, but also, and more importantly, because it forms part of an approach, that of the theory of the actor-network or translation, which is able to make direct inferences about the dynamics of scientific development.

PROSPECTS FOR SOCIAL THEORY

The approach that has been developed in this book has implications not only for science and science policy and the role of science and technology in society, but also for our understanding of the nature of social science inquiry, the relationship between the 'macro' and the 'micro', and the analysis of power and long distance social control.

Consider first the nature of social science inquiry. Here the social investigator is no longer considered to have privileged access to social reality (see Callon and Latour, 1985). Thus sociologists of science, even if they are relatively modest in their claims about the substance of science, often allow themselves to make statements about social interests and group alignments as if their knowledge of these were neither limited nor flawed. By contrast, our claims about the nature of social structure are restricted and, when we make such suggestions, it has to be understood that we believe ourselves to be doing exactly the same as any other individual: we are building a world and then attempting to convert that world into reality by enrolling others in it. The reader is thus being asked to believe in our co-word maps, and we are using the same heterogeneous means of persuasion that we have claimed to be available to the scientists that we study. However this does not mean, as some might imagine, that our position is self-refuting. We are like everyone else. All that we can do is to make arguments that we hope will recommend themselves to some of our readers.[1]

As a complement to this point, we note that there is no one real and definitive pattern of scientific development or of the structure of society that is, so to speak, waiting in the wings to be discovered. Rather, there is a multitude of perspectives, each struggling to extend its scope and its influence (Latour and Strum, 1985). The fact that for a given area of science and technology different co-word maps can be drawn depending on the choice of databases, indices and thresholds, is thus not a shortcoming of the method but rather a sign of its strength. It reflects the essential plurality of perspectives (which are, however, not equivalent: some are, as a matter of fact, stronger than others).

The particular advantage of our method is that, instead of trying to impose another strong conception of the nature of social reality, we rather study the way in which actors attempt to impose worlds upon one another. 'Follow the actors' – this is the methodological cornerstone of the approach.

A corollary of this method is that the distinction between macro- and micro-analysis disappears. In social science so-called large-scale issues are usually discussed with the help of notions such as structure and system, while the small scale is analysed in terms of interactions or tasks. However, within actor-network theory that which is large is that which has successfully translated others and has therefore grown. Since size is nothing more than the end-product of processes of translation, the need for two analytical vocabularies is thus avoided. Indeed, to make use of a separate vocabulary for the large tends to conceal both the processes by which growth occurs, and the uncertainties that are involved in maintaining power and size. In addition, it reifies the status of the large, and makes it appear as if the latter could never decrease in size and become weak. We believe that the social theorist has no reason to aid those who happen to be powerful. Even-handedness is just as appropriate here as anywhere else. The adoption of a single vocabulary of analysis, that of translation, helps to remind us that if certain actors are more powerful than others then this is an achievement and not something with which they are endowed by nature.[2]

The importance of the actor-network approach is not simply methodological. One may also ask more theoretical questions about the internal structure of actor-networks. In particular one may ask about 'centres of translation' – that is, the places where translation strategies are evolved, where attempts are made to control the diverse elements that make up the actor-network, and calculations about the return from different strategies are made. In Chapter 2 we saw the way in which Electricité de France, in attempting to build a world full of electric vehicles, turned itself, at least for a time, into such a centre of translation and thus became the spokesman for a range of diverse natural and social entities.[3] How do such centres of translation achieve their status? How are they elected to be spokesman? Why do some find that others are speaking on their behalf? The answers to these questions will depend on both empirical and theoretical work.

Similar questions can also be raised in a more historical context: Are there dominant types of centres of translation in certain societies at certain times? Is it possible to trace shifts in the types of centres that are dominant? In Chapter 1 we made an argument of this kind by identifying laboratories (and their analogues) as the strategic loci in contemporary society. Support for this argument can be found in studies of Pasteur's microbiological work (Latour, 1984) and of the French electric vehicle controversy (Chapter 2 of this volume). However, it needs to be filled out by considering, for instance, how texts were able, historically, to become emissaries that are durable, transportable and forceful, and therefore crucial agents of social control. The advent of the printing press is clearly important here (Latour, 1985, following Eisenstein, 1979). But what about other kinds of emissaries such as people and devices? Does the advent of drill (Foucault, 1979; McNeill, 1982) mark a similar breakthrough in the creation of disciplined bodies? Would the invention of manufacture (see, e.g. Smith, 1977) count as an equivalent step in the history of machines and devices? And may the historical dominance of the West be understood in terms of a combination of such revolutions in the means of translation (Law, 1985d)?

Such questions, at present speculative, lead to further reappraisals of traditional social-theoretic issues. For instance, may classical bureaucracies be seen as functioning primarily by a combination of drilled people and texts, and armies in terms of a people/device duo? And may we expect further revolutions in the means of translation, possibly in relation to what is sometimes called the information society? Such questions relate to issues that have been discussed by writers from Weber to Habermas and Marcuse: the inescapable rationality of bureaucratic administration and the possibly dehumanising power of technology. But looked at from the standpoint of the present approach, these issues now appear as instances of social control – of long distance translation. And this approach implies that such control is not monolithic. Rather, there is a wide range of struggling actors and there are periodic changes in both the means of control and the strategic loci.

There may, indeed, be secular trends. However, the actor-network approach does not share the implicit or explicit pessimism of the writers mentioned above. Particular forms of domination are clearly reversible, and the means of world-building are not necessarily restricted to the few.

The social investigator who follows the actors and describes the strategies of the great will, like Machiavelli, also offer aid to those who are dominated.

As this book gives witness, the actor-network approach has been developed in the context of the sociology of science and technology. However, as we have briefly intimated, the theory is potentially general in nature and has implications for both theories of power and practical attempts at social control. It is in this context that the importance of texts can be seen. Their influence cannot be understood apart from the circumstances in which they act. Nevertheless they constitute one of the most important contemporary means of long distance social control.

NOTES

1. For a discussion of this type of argument see Law and Lodge (1984, pp. 223–9).
2. These arguments have been more fully developed elsewhere, see Callon and Latour (1981) and Law (1984a).
3. For a development of these aspects of translation, see Callon (1985).

Bibliography

Adler, Reid G. (1984) 'Biotechnology as an Intellectual Property', *Science*, vol. 224 (27 April) pp. 357–63.

Allen, T. J. (1984) *Managing the Flow of Technology* (Cambridge, Mass.: MIT Press).

Anderson, P., P. Miller, and P. Brockman (1983) *New Essays in Scientific and Technical Communication* (Farmingdale, New York: Baywood).

Anon (1978) *Pour une aquaculture autogérée en Bretagne*, mimeo (Brest: Bretagne Documentation).

Anon (1981) *Programme de décentralisation du laboratoire de physiologie des poissons de Jouey-en-Josas à Rennes*, mimeo (Paris: INRA).

Anon (1982) *La truite elevée en mer: comment la vendre*, mimeo (Rennes: GRESA–ANO).

Aron, Raymond (1972) 'Preface' to Machiavelli's *Le Prince* (Paris: Livre de Poche).

Barnes, Barry (1977) *Interests and the Growth of Knowledge* (London: Routledge & Kegan Paul).

Bastide, Françoise (1979) 'Le foie lavé, approche sémiotique d'une texte de sciences expérimentales', *Documents de Recherche*, Groupe de Recherches Sémiolinguistiques, vol. 1, no. 7, pp. 9–34.

Bastide, Françoise (1980) 'Remarque sur le faire interprétif du chercheur' *Le Bulletin du Groupe de Recherches Sémiolinguistiques*, vol. 1, no. 7, pp. 26–36.

Bastide, Françoise (1981a) 'Variation sur la vérédiction', *Le Bulletin du Groupe de Recherches Sémiolinguistiques*, vol. 17, pp. 20–3.

Bastide, Françoise (1981b) 'Le sentier et la cascade; deux figures spontanément aspectualisées' *Le Bulletin du Groupe de Recherches Sémiolinguistiques*, pp. 16–26.

Bastide, Françoise (1981c) 'La démonstration, analyse de la structure actantielle du faire-croire', *Documents de Recherche*, vol. 3, pp. 28–40.

Bazerman, Charles (1981) 'What Written Knowledge Does: Three Examples of Academic Discourses', *Philosophy of the Social Sciences*, vol. 11, pp. 361–7.

Bazerman, Charles (1984) 'Modern Evolution of the Experimental Report in Physics: Spectroscopic Articles in *Physical Review*, 1893–1980', *Social Studies of Science*, vol. 14, pp. 163–96.

Benzecri, J. P. *et al.* (1973) *L'analyse des données, Tome 1, La taxonomie, Tome 2, L'analyse des correspondences* (Paris: Dunod).

Bijker, Wiebe, Trevor Pinch and Thomas P. Hughes (1986) *New Directions in the Social Study of Technology* (Cambridge, Mass.: MIT Press).

Bloor, David C. (1976) *Knowledge and Social Imagery* (London: Routledge & Kegan Paul).

Böhme, Gernot, Wolfgang van den Daele, Rainer Hohlfeld, Wolfgang Krohn and Wolf Schäfer (1983) *Finalization in Science: the Social Orientation of Scientific Progress* (Dordrecht, Boston and Lancaster: Reidel).

Bouzat, J. M. and P. A. Grislain (1979) *Les actions concertées chimie macromoleculaire*, final report to the DGRST (Paris: Centre de Sociologie de l'Innovation, École Nationale Supérieure des Mines).

Bukharin, N.I. *et al.* (1971) *Science at the Crossroads*, papers from the Second International Congress of the History of Science and Technology, 1931, reprinted (London: Frank Cass).

Callon, Michel (1975) 'L'opération de traduction comme relation symbolique', in Roqueplo, P. (1975) pp. 105–41.

Callon, Michel (1980a) 'The State and Technical Innovation: a Case Study of the Electric Vehicle in France', *Research Policy*, vol. 9, pp. 358–76.

Callon, Michel (1980b) 'Struggles and Negotiations to Define What Is Problematic and What Is Not. The Socio-logic of Translation', in Knorr-Cetina, Karin D., Roger Krohn and Richard D. Whitley (1980) pp. 197–219.

Callon, Michel (1981a) 'Pour une sociologie des controverses technologiques', *Fundamenta Scientiae*, vol. 2 (3/4), pp. 381–99.

Callon, Michel (1981b) 'Boîtes noires et opérations de traduction', *Economie et Humanisme*, vol. 262 (Nov.) pp. 53–59.

Callon, Michel (1983) 'L'agonie d'un laboratoire saisi par le démon de la technologie', mimeo (Paris: Centre de Sociologie de l'Innovation, École Nationale Supérieure des Mines).

Callon, Michel (1985) 'The Domestication of the Scallops and Fishermen of St Brieuc Bay: Some Elements of a Sociology of Translation', in Law, John (1985c), forthcoming.

Callon, Michel, Jean-Pierre Courtial and William A. Turner (1979) *Les actions concertées chimie macromoleculaire. Sociologie d'une agence de traduction*, final report to the DGRST (Paris: Centre de Sociologie de l'Innovation, École Nationale Supérieure des Mines).

Callon, Michel, Jean-Pierre Courtial, William A. Turner and Serge Bauin (1983) 'From Translations to Problematic Networks: An Introduction to Co-word Analysis', *Social Science Information*, vol. 22, pp. 199–235.

Callon, Michel and Bruno Latour (1981) 'Unscrewing the Big Leviathan: How Actors Macrostructure Reality and Sociologists Help Them to Do So', in Knorr-Cetina, Karin D. and Aaron Cicourel (1981), pp. 277–303.

Callon, Michel and Bruno Latour (1985) 'Pour une sociologie relativement exacte', *Année Sociologique*, vol. 86, forthcoming.

Callon, Michel and John Law (1982) 'On Interests and their Transformation', *Social Studies of Science*, vol. 12, pp. 615–25.

Cole, J. and S. Cole (1973) *Social Stratification in Science* (University of Chicago Press).

Collins, H.M. (1975) 'The Seven Sexes: A Study in the Sociology of a Phenomenon, or the Replication of Experiments in Physics' *Sociology*, vol. 9, pp. 205–24.

Collins, H.M. and Trevor Pinch (1982) *Frames of Meaning: The Social Construction of Extraordinary Science* (London: Routledge & Kegan Paul).

Commissioner of Patents and Trademarks (1983) 'Automating the Patent and Trademark Office', Report to Congress (Washington: Patent and Trademark Office).

Courtial, Jean-Pierre and Michel Callon (1985) 'Is Indexing Trustworthy? Classification of Articles Through Co-word Analyses', *Journal of Information Science*, vol. 9, forthcoming.

Crane, Diana (1972) *Invisible Colleges* (University of Chicago Press).

Ducasse, R. *et al.* (1985) 'Journal-review de science de l'information: essai de définition et projet de réalisation d'un nouveau système de communication des informations produites par et pour les sciences et les techniques', working paper (Paris: Ministry of Education).

Edge, David O. (1979) 'Quantitative Measures of Communication in Science: a Critical Review', *History of Science*, vol. 17, pp. 102–34.

Eisenstein, Elizabeth (1979) *The Printing Press as an Agent of Social Change: Communications and Cultural Transformations in Early-Modern Europe* (Cambridge University Press).

Elias, Norbert, Herminio Martins and Richard D. Whitley (eds) (1982) *Scientific Establishments and Hierarchies, Sociology of the Sciences Yearbook* vol. 6 (Dordrecht and Boston: Reidel).

Elkana, Y., J. Lederberg, R. K. Merton, A. Thackray and H. Zuckerman (eds) (1978) *Toward a Metric of Science: the Advent of Science Indicators* (New York: John Wiley).

Ellis, P., G. Hepburn and C. Oppenheim (1978) 'Studies on Patent Citation Networks', *Journal of Documentation*, vol. 134, pp. 12–20.

Feyerabend, Paul K. (1975) *Against Method* (London: New Left Books).

Foucault, Michel (1979) *Discipline and Punish: the Birth of the Prison* (Harmondsworth: Penguin).

Freeman, Chris (1983) *The Economics of Industrial Innovation* (Harmondsworth: Penguin).

Garfield, Eugene (1979) *Citation Indexing, its Theory and Application in Science, Technology and Humanities* (New York: John Wiley).

Garfield, Eugene, M. V. Malin and Henry Small (1978) 'Citation Data as Science Indicators', in Elkana, Y. *et al.* (1978) pp. 179–207.

Garfinkel, Harold (1967) *Studies in Ethnomethodology* (New Jersey: Prentice Hall).

Gerth, H.H. and C. Wright Mills (eds) (1948) *From Max Weber, Essays in Sociology* (London: Routledge & Kegan Paul).

Gilbert, G. Nigel and Michael Mulkay (1984) *Opening Pandora's Box: A Sociological Analysis of Scientists' Discourse* (Cambridge University Press).

Gilbert, G. Nigel and Steven Woolgar (1974) 'The Quantitative Study of Science: An Evaluation of the Literature', *Science Studies*, vol. 4, pp. 279–94.

Goldstein, Jorge A. (1982). 'A Footnote to the Cohen-Boyer Patent and other Musings', *Recombinant DNA Technical Bulletin* (Washington National Institutes of Health), vol. 5, pp. 180–8.

Goody, Jack (1977) *The Domestication of the Savage Mind* (Cambridge University Press).

Goody, Jack (1980) *La Raison Graphique* (Paris: Minuit).

Gondran, M. and M. Minoux (1979) *Graphes et algorithmes,* Collection de la Direction des Études et Recherches d'Electricité de France (Paris: Eyrolles).

Gopnik, M. (1972) *Linguistic Structure in Scientific Texts* (Paris: Mouton).

Granger, G.G. (1968) *Essai d'une philosophie du style* (Paris: Armand Colin).

Griffith, Belver C., Henry Small, J. A. Stonehill and S. Dey (1974) 'The Structure of Scientific Literatures II: Toward a Macro- and Micro-Structure

for Science', *Science Studies* vol. 4, pp. 339–65.
Grmek, M. (1973). *Raisonnement expérimental et recherche toxicologique chez Claude Bernard* (Geneva: Droz).
Gusfield, J. (1976) 'The Literary Rhetoric of Science: Comedy and Pathos in Drinking Driver Research', *American Sociological Review*, vol. 41, pp. 16–34.
Haberkorn, J-P. (1975) 'Les promesses de l'aquaculture', *Problèmes economiques*, pp. 8–10 (Paris: la Documentation Française).
Habermas, Jurgen (1971) *Towards a Rational Society* (London: Heinemann).
Healey, Peter and Harry Rothman (1984) 'Evaluative Summary Report of the ABRC Science Policy Study, 1983/84', mimeo (London and Birmingham: ESRC/Aston University).
Hesse, Mary B. (1974) *The Structure of Scientific Inference* (London: Macmillan).
Hessen, Boris (1971). 'The Social and Economic Roots of Newton's *Principia*', in Bukharin, N. I. *et al.* (1971).
Hill, M. (1974) 'Correspondence analysis: a neglected multivariate method', *J. Appl. Statistics*, vol. 23, pp. 340–54.
Hughes, Thomas P. (1983) *Networks of Power: Electrification in Western Society, 1880–1930* (Baltimore: Johns Hopkins University Press).
Huni, A. and V. Buss (1982) 'Patent Protection in the Field of Genetic Engineering', *Industrial Property* (Dec.) pp. 356–68.
Ivins, W. (1938) *On the Rationalization of Sight*, reprinted 1975 (New York: Plenum Press).
Ivins, W. (1973) *Prints and Visual Communication*, reprinted 1979 (Cambridge, Mass.: MIT Press).
Jacobi, D. (1982) *La diffusion des connaissances scientifiques: stratégies des chercheurs auteurs d'articles dans une revue de vulgarisation*, Thèse de troisième cycle, University of Besançon.
Keriguy, J. (1984) 'L'accès à la littérature grise: une contribution du CDST', *Bulletin des Bibliotheques de France*, vol. 29, pp. 138–43.
Knorr-Cetina, Karin D. (1981) *The Manufacture of Knowledge: An Essay on the Constructivist and Contextual Nature of Science* (Oxford and New York: Pergamon Press).
Knorr-Cetina, Karin D. and Aaron Cicourel (eds) (1981) *Advances in Social Theory and Methodology: Toward an Integration of Micro- and Macro-Sociologies* (London: Routledge & Kegan Paul).
Knorr-Cetina, Karin D., Roger Krohn and Richard D. Whitley (eds) (1980) *The Social Process of Scientific Investigation, Sociology of the Sciences Yearbook*, vol. 4 (Dordrecht and Boston: Reidel).
Knorr-Cetina, Karin D. and Michael Mulkay (eds) (1983) *Science Observed: Perspectives on the Social Study of Science* (London and Beverly Hills: Sage).
Koyré, A. (1968) *Etudes Newtoniennes* (Paris: Gallimard).
Kuhn, Thomas, S. (1970) *The Structure of Scientific Revolutions* (Chicago University Press).
Lakatos, Imre (1970) 'Falsification and the Methodology of Scientific Research Programmes', in I. Lakatos and A. Musgrave (1970) pp. 91–195.
Lakatos, Imre and Alan Musgrave (eds) (1970) *Criticism and the Growth of Knowledge* (Cambridge University Press).
Lancaster, F.W. (1982) 'The Evolving Paperless Society', *International Forum on Information and Documentation*, vol. 7, pp. 3–10.

Latour, Bruno (1983) 'Give me a Laboratory and I will Raise the World', in Knorr-Cetina, Karin D. and Michael Mulkay (1983) pp. 141–70.

Latour, Bruno (1984) *Les microbes, guerre et paix* followed by *Irréductions* (Paris: Metailié).

Labour, Bruno (1985) 'Visualisation and Cognition', *Knowledge and Society*, forthcoming.

Latour, Bruno, and P. Fabbri (1977) 'La rhétorique de la science: pouvoir et devoir dans un article de la science exacte', *Actes de la Recherche en Sciences Sociales*, vol. 13, pp. 81–95.

Latour, Bruno and Shirley Strum (1985) 'The Meanings of Social', in Law, John (1985c) forthcoming.

Latour, Bruno and Steven Woolgar (1979) *Laboratory Life: the Social Construction of Scientific Facts* (London and Beverly Hills: Sage).

Law, John (1983) 'Enrôlement et contre-enrôlement: les luttes pour la publication d'un article scientifique', *Social Science Information*, vol. 22, pp. 237–51.

Law, John (1984a) 'How Much of Society Can the Sociologist Digest at One Sitting? The "Macro" and the "Micro" Revisited for the Case of Fast Food', *Studies in Symbolic Interaction*, vol. 5, pp. 171–96.

Law, John (1984b) 'A propos des tactiques du control social: une introduction à la théorie de l'acteur-réseau, *La Légitimité Scientifique; Cahiers Science, Technologie, Sociëté*, vol. 4, pp. 106–26 (Paris: CNRS).

Law, John (1985a) 'A propos des mots et des autres alliés', *Culture Technique*, forthcoming.

Law, John (1985b) 'Technology, Closure and Heterogeneous Engineering: The Case of the Portuguese Expansion' in Bijker, Wiebe, Trevor Pinch and Thomas Hughes (1986) forthcoming.

Law, John (ed.) (1985c) *Power, Action and Belief: The New Sociology of Knowledge, Sociological Review Monograph 32* (Keele and London: University of Keele and Routledge & Kegan Paul) forthcoming.

Law, John (1985d) 'On the Methods of Long Distance Control: Vessels, Navigation and the Portuguese Route to India', in Law, John (1985c) forthcoming.

Law, John, and Peter Lodge (1984) *Science for Social Scientists* (London: Macmillan).

Law, John and Rob Williams (1981) 'Social Structure and Laboratory Practice', paper presented to the Conference on Communication and Scientific Research, Vancouver, Simon Fraser University.

Law, John and Rob Williams (1982) 'Putting Facts Together: A Study in Scientific Persuasion', *Social Studies of Science*, vol. 12, pp. 535–58.

Lelarge, S. (1980) *L'aquaculture n'est pas une île* (Bordeaux: ADA).

Levidov, Les and Bob Young (eds) (1981) *Science, Technology and the Labour Process* (London: CSE Books).

Lofflet-Laurian, A-M. (1980) 'L'expression du locuteur dans les discours scientifiques, "Je", "Nous", et "On" dans quelques textes de chimie et de physique', *Revue de Linguistique Romane*, vol. 17, pp. 135–57.

Lynch, Michael (1982) 'Technical Work and Critical Inquiry: Investigations in a Scientific Laboratory', *Social Studies of Science*, vol. 12, pp. 499–534.

Lynch, Michael (1985) *Art and Artefact in Laboratory Science: A Study of Shop Work and Shop Talk in a Research Laboratory* (London: Routledge & Kegan Paul).

Machiavelli, Niccolo (1950) *Discourses* (London: Routledge & Kegan Paul).
Machiavelli, Niccolo (1961) *The Prince* (Harmondsworth: Penguin).
Machlup, Fritz (1972) *The Production and Distribution of Knowledge in the United States* (Princeton University Press).
MacKenzie, Donald (1978) 'Statistical Theory and Social Interests: A Case Study', *Social Studies of Science*, vol. 8, pp. 35–83.
McNeill, William H. (1982) *The Pursuit of Power: Technology, Armed Force and Society Since AD 1000* (Oxford: Blackwell).
Martin, B.R. and J. Irvine (1983) 'Assessing Basic Research: Some Partial Indicators of Scientific Progress in Radio Astronomy'. *Research Policy*, vol. 12, pp. 61–90.
Meadows, A.J. (1974) *Communication in Science* (London: Butterworth).
Meadows, A.J. (ed.) (1980) *The Development of Scientific Publishing in Europe* (Amsterdam: Elsevier).
Medawar, Peter (1964) 'Is the Scientific Paper fraudulent?', *Saturday Review*, vol. 47 (August) pp. 42–3.
Menard, Henry (1971) *Science: Growth and Change* (Cambridge, Mass.: Harvard University Press).
Merton, Robert K. (1973) in Norman W. Storer (ed.) *The Sociology of Science: Theoretical and Empirical Investigations* (Chicago University Press).
Mullins, N.C. (1972) 'A Model for the Development of a Scientific Specialty: The Phage Group and the Origins of Molecular Biology', *Minerva*, vol. 10, pp. 51–82.
Narin, F. *et al.* (1976) *Evaluative Bibliometrics. The Use of Publication and Citation Analysis in the Evaluation of Scientific Activity* (Cherry Hill, New Jersey: Computer Horizons).
National Science Board (1973, 1975, 1977, 1979, 1981) *Science Indicators* (Washington DC: US Government Printing Office).
OECD (1980) *The Measurement of the Output of R and D Activities: The 1980 Conference on Science and Technology Indicators*, (Paris: OECD).
OECD (1982) *Workship on Patent Innovation Statistics*, (Paris: OECD).
Ong, Walter (1982) *Orality and Literacy: the Technological Ring of the Word* (London: Methuen).
Pavitt, Keith (1978) 'Using Patent Statistics in *Science Indicators*', in *The Meaning of Patent Statistics*, prepared for the National Science Foundation (Washington DC: NSF).
Pavitt, Keith (1982) 'R and D, Patenting and Innovative Activities: a Statistical Exploration', *Research Policy*, vol. 11, pp. 33–51.
Perelman, Charles (1970) *Traité de l'argumentation, la nouvelle rhétorique* (Brussels: Editions de l'Institut de Sociologie).
Pinch, Trevor and Wiebe Bijker (1984) 'The Social Construction of Facts and Artefacts: Or How the Sociology of Science and the Sociology of Technology Might Benefit Each Other', *Social Studies of Science*, vol. 14, pp. 399–441.
Popper, Karl (1959) *The Logic of Scientific Discovery* (London: Hutchinson).
Price, Derek de Solla (1963) *Little Science, Big Science* (New York: Columbia University Press).
Rip, Arie (1981) 'A Cognitive Approach to Science Policy', *Research Policy*, vol. 10, pp. 294–311.
Rip, Arie and Jean-Piere Courtial (1984) 'Co-word Maps of Biotechnology: An Example of Cognitive Scientometrics', *Scientometrics*, vol. 6, pp. 381–400.

Roqueplo, P. (ed.) (1975) *L'incidence des rapports sociaux sur la science* (Paris: Cordes).

Rousseau, Jean-Jacques (1968) *The Social Contract* (Harmondsworth: Penguin).

Schmookler, Jacob (1966) *Invention and Economic Growth* (Cambridge, Mass.: Harvard University Press).

Schumpeter, Joseph A. (1926) *The Theory of Economic Development* (Oxford University Press).

Shapin, Steven (1979) 'The Politics of Observation: Cerebral Anatomy and Social Interests in the Edinburgh Phrenology Disputes', in Wallis, Roy G. (ed.) (1979) pp. 139–78.

Small, Henry G. (1977) 'A Co-citation Model of a Scientific Specialty: a Longitudinal Study of Collagen Research' *Social Studies of Science*, vol. 7, pp. 139–66.

Small, Henry G. and E. Greenlee (1980a) 'Citation Context Analysis and the Structure of Paradigms', *The Journal of Documentation*, vol. 36, pp. 183–96.

Small, Henry G, and E. Greenlee (1980b) 'Citation Context Analysis of a Co-citation Cluster: Recombinant DNA' *Scientometrics*, vol. 2, pp. 277–301.

Small, Henry G. and Belver C. Griffith (1974) 'The Structure of Scientific Literatures I and II: Identifying and Graphing Specialties', *Science Studies*, vol. 4, pp. 17–40 and 339–65.

Smith, Merrit Roe (1977) *Harper's Ferry Armory and the New Technology: The Challenge of Change* (Ithaca and London: Cornell University Press).

Star, Susan Leigh (1983) 'Simplification in Scientific Work: An Example from Neuroscience Research', *Social Studies of Science*, vol. 13, pp. 205–28.

Thackray, Arnold (1972) *John Dalton: Critical Assessments of his Life and Science* (Cambridge, Mass.: Harvard University Press).

Toulmin, S.E. (1972) *Human Understanding, Part 1* (Oxford University Press).

Touraine, Alain (1973) *Production de la Société* (Paris: Le Seuil).

Van den Belt, Henk and Arie Rip (1984) *Technologie-ontwikkeling: het Nelson-Winter/Dosi Model* (University of Leiden).

Wallis, Roy G. (ed.) (1979) *On the Margins of Science: The Social Construction of Rejected Knowledge, Sociological Review Monograph 27*, University of Keele.

Weber, Max (1948) 'Politics As a Vocation', in Gerth, H. and C.W. Mills (1948) pp. 77–128.

Weingart, Peter (1982) 'The Scientific Power Elite – a Chimera. The De-institutionalization and Politicization of Science' in Elias, Norbert, Herminio Martins and Richard D. Whitley (eds) (1982) pp. 71–87.

Williams, Rob (1984) *The Matter-of-Fact in Science: A Sociological Ethnography of a Biochemistry Laboratory*, PhD thesis, University of Keele.

Williams, Rob and John Law (1980) 'Beyond the Bounds of Credibility', *Fundamenta Scientiae*, vol. 1, pp. 295–315.

Woolgar, Steven (1976) 'Writing an Intellectual History of Scientific Development: The Use of Discovery Accounts' *Social Studies of Science*, vol. 6, pp. 395–422.

Woolgar, Steven (1980) 'Discovery: Logic and Sequence in a Scientific Text', in Knorr-Cetina, Karin D., Roger Krohn and Richard Whitley (eds.) (1980) pp. 239–68.

Yearley, Steven (1981) 'Textual Persuasion: the Role of Social Accounting in the

Construction of Scientific Arguments', *Philosophy of the Social Sciences*, vol. 11, pp. 409–35.

Yoxen, Edward (1981) 'Life as a Productive Force: Capitalizing upon Science and Technology of Molecular Biology', in Levidov and Young (1981) pp. 66–122.

Index

Page references in italics indicate pages where a definition or systematic introduction occurs.

accumulator, 86–9, *see also* electrochemical batteries
act(ion) at a distance, 10, 27, 50, 84
actor, xvi, 4, 15n, 29, 95, 98, 107, 110, 144, 184, 187, 192, 203, 216, 223–6, 228, 230
actor-network, *xvi*, *28–34*, 42, 46, 50, 70–1, 81, 105–8, 122, 211, 221, 224, 228–30
actor-world, *xvi*, *20–4*, 26, 28, 30–3, 67, 77, 81, 84–5, 89–90, 107, 111, 226
Adler, R. G., 93, 99n
aggregation, 12–3, 27, 103–4, 107, 110–1, 147, 160, 223, 225
Allen, T. J., 146
analyst, 10, 176, 215, 225
Anderson, P., 66n
anti-reductionism, 7
aquaculture, 120–2, 124–41
Aron, R., 6
article, *see* scientific article

Barnes, B., 8
Bastide, F., 62, 66n
Bauin, S., 105, 120, 123n
Bazerman, C., 66n
Benzecri, J. P., 190
bibliometrics, 112
Bijker, W., 20
biotechnology, 93–6, 116, 163–4, 184
black box, *xvi*, 29, 31, 34, 40, 90, 92, 96, 115–16, 123n, 139, 223
Bloor, D., 8
Boehme, G., 103
Bouzat, J. M., 144
bridge of interests, 88, 91
Bukharin, N., 7
Buss, V., 99n

Callon, M., 20, 24, 34n, 81n, 82n, 88, 91, 105, 120, 123n, 143, 192, 203, 230n
cancer, 38, 49, 69, 74, 80
CDST (Scientific and Technical Documentation Centre), 111, 117
central pole, *114–16*, 119, 125–31, 135, 137–9, 156–9, 164–8, 187–8n, 193–4, 201, 205, 216, 226
centre of translation, *see* translation centre
chain of translation, 119, 149, 155, 179, 226
of words, 164, 166, 176, 191–2, 197, 203–4
co-citation analysis, 104, 123n, 140, 210n, 211, 226
coefficient, *see* inclusion coefficient, Jaccard index, proximity coefficient, statistical coefficient
Cole, J., 145
Collins, H. M., 103
consolidation, 115–16, 186–7
co-occurrence, 108–9, *112*, 114, 195–6, 225
co-occurrence matrix, *112*, 189, 227
Courtial, J. P., 105, 111, 120, 123n, 143, 192–3, 203
co-word analysis, 3, 81, 104, 107–8, *110–17*, 122, 140, 144, 160–1, 163–4, 187–9, 191–3, 198–9, 203, 209, 211–12, 214–7, 221, 225–6
co-word map, 117, 124, 224, 226, 228
Crane, D., 145

database, 125, 144–5, 163, 187, 189, 193, 199, 211–6, 225, 228
dietary fibre, 111, 117–20, 189, 194, 198, 200–3, 206–9, 215
displace(ment), xvii, 4, 6, 27–8, 85, 91, 105–6
DIVEMA, 12, 69, 72–9, 82n, 83, 106
documentation service, 13, 111, 199, 209
document retrieval, 176, 211–12, 215–16
Ducasse, R., 213

durability, 13, 23, 28, 32, 49, 69, 77, 80, 105, 224, 229
dynamics
  of science/technology, 3, 12, 14, 96, 103–4, 107–8, 122, 187, 217, 227
  of (scientific) fields, 179, 217

EDF (Electricité de France), 21–32, 39, 86–9, 104–5, 228
Edge, D. O., 104
Eisenstein, E., 15n, 66n, 229
electric vehicle, 21–3, 71, 78, 86–8, 105, 228–9
electrochemical batteries, 21–3, 27, *see also* accumulator
Elkana, Y., 12, 123n
Ellis, P., 123n
enrol(ment), *xvi*, 24–5, 68, 71, 77, 80, 92, 105, 222, 227
entities, 14, 22, 25–7, 28–30, 32–4, 71, 105, 222, 224, 228
entrepreneur(ial), 9, 38, 104
equivalence, xvii, 58, 75, 78, 80, 105–10, 112, 114, 122, 190
ethnography, 5, 6, 9, 11–12, 98
experiment, 36–9, 57, 75, 77–8

fence of interest, 92, 95, 97
Feyerabend, P., 66n
firm, industrial, 98, 142, 155–7, 160, 184–7, 216
force
  of network, 31, 34
  of science, 3, 6, 14, 79, *see also* power of science
  of texts, 10, 51, *68*, 72, 96–7
Foucault, M., 229
Freeman, C., 123n, 188n
fuel cell, 21–3, 29–32, 86–9, 98n, 105, 222
funding applications, 84–91, 145–8, 157, 160
funnel of interest, *xvi*, *77–80*, 82n, 90

Garfield, E., 15n, 104, 123n
Garfinkel, H., 111
Gilbert, G. N. 66n, 123n
Goldstein, J. A., 95, 99n
Gondran, M., 141n, 192
Goody, J., 15n, 66n
Gopnik, M., 66n
Granger, G. G., 66n
graph, 106–8, 112–13, 141n, 148, 168, 171, 176, 192–5, 197, 199, 203
Griffith, B. C., 15n
Grislain, P. A., 144
Grmek, M., 66n
Gusfield, J., 66n

Habermas, J., 7, 229
Healey, P., 211, 216
Hepburn, G., 123n
Hesse, M., 34n, 103, 115
Hessen, B., 20
heterogeneity, 121–2, 224, 226
heterogeneous elements/entities, 9, 15, 23, 31, *38–42*, 48–9, 67, 72, 79, 81, 105–6, 191, 227
hierarchical position, 114, *see also* rank
  relations, 113, 115
  structure, 147
hierarchisation, 112, 115, 139
hierarchy, 110, 112, 116, 119, 122, 141n
hierarchy of problematisation, 87–90, 112
Hill, M., 190
Hohlfeld, R., 103
Hughes, T., 21
Huni, A., 99n

inclusion coefficient/index, *112–14*, 131, 169–70, 192–3, 195–6
inclusion map, *113–14*, *116–19*, 122, 125–32, 135, 137–8, 156, 164–70, 176, 179, 186–7, 201, 203, 205–6, 215
index, *see* inclusion index, Jaccard index, proximity index, statistical coefficient
indexation, 110, 130–1, 146, 163–4, 188, 190, 198–9, 201, 205–6, 226
indexer, 111, 130–1, 191, 198–9, 201, 203, 206, 212, 215, 226
indexing service, 111, 215
industry, 3, 4, 108, 139, 145–7, 149, 151–7, 159–61, *see also* firm
inscriptions, *9–10*, 27, 46, 59
inscription device, 54
interaction, 12, 34, 107–8, 161
interessement, *xvii*, *70–5*, 77, 80, 82n, 85, 87, 91, 96, 98n
interests, 7, 25, 34, 67–9, 71, 75, 77, 83–6, 89, 93, 108, 135, 145–7, 149, 152–7, 159, 161–2, 186–7, 192
interesting, 43, 69, 71, 96, 226–7
interpretation, 52
Irvine, J., 142
itinerary, xvii, 27, 90–33, 222

Jaccard index, 193
Jacobi, D., 66n
journal, *see* scientific journal

junction word, 122
juxtaposition, xvi, 9, 12, 30, 39, 42, 46, 49, 68, 70, 105, 226

key-words, 91, 97, 98n, 110–12, 117, 122, 125, 147, 157, 185–6, 189–92, 195, 198–9, 205–7, 210n, 211
Knorr-Cetina, K. D., 9, 20, 52, 66n
Koyré, A., 66n
Krohn, W., 103
Kuhn, T. S., 103, 116

laboratory, 4, 8, 14–15, 19–20, 27, 35–40, 48–50, 58, 61, 67, 72, 77–8, 81, 84–90, 98, 108, 116, 142, 145, 151, 207–9, 216, 221–4, 229
Lakatos, I., 103
Lancaster, F. W., 212
Latour, B., 4, 9, 15n, 20–1, 27, 66n, 81n, 82n, 90, 103, 222, 228–9, 230n
Law, J., 9, 20, 27, 33, 34n, 52, 66n, 81n, 82n, 106, 123n, 223, 229, 230n
Levidov, L., 7
linkages, 109–10, 122, 197
Lodge, P., 230n
Lofflet-Laurian, A-M., 66n
Lynch, M., 9, 52, 66n

Machiavelli, N., 5–6, 223–4, 230
Machlup, F., 123n
MacKenzie, D., 8, 20, 103
McNeill, W. H., 229
macromolecular chemistry, 142–5, 156, 160, 162, 191
maps, 13, 90–1, 110, 114, 131, 151, 156, 164, 168, 203, 215–16, 225–7, *see also* inclusion map, proximity map
mapping, 11–13, 91–2, 98n, 111
Marcuse, H., 229
Martin, B. R., 142
Marx, K., 7
Meadows, A. J., 66n, 145
Medawar, P., 66n
mediator words, *114*, 120, 125–31, 164–6, 179, 201, 203, 205–7, 209, 210n, 226
Menard, H., 123n
Merton, R. K., 7
Minoux, M., 141n, 192
mobilisation, 31, 71, 77, 84, 97–8, 224
mobility, 10
Mulkay, M. J., 66n
Mullins, N., 15n, 103
multidimensional anaysis, 190–2

Narin, F., 123n
network, xvi, 30–2, 42, 74, 81, 87, 106, 108–10, 113, 123n, 133, 144–5, 148–9, 161, 162n, 204–5, 213, 226
non-reductionism, 8

obligatory point of passage, xvii, 26–7, 75, 81, 84, 90, 94, 97–8, 103, 105, 122, 188, 222
observer, 53–4
occurrence, 108, 110, *112–13*, 168
OECD (Organisation for Economic Cooperation and Development), 123n, 142, 188n
Ong, W., 66n
Oppenheim, C., 123n

palace, 4, 5, 98, 222, 224
paper, *see* scientific paper
passageway, 26, 84, 93–4, 96, 119, 122, 160–1, 224, *see also* obligatory point of passage
  on co-word maps, 126, 149, 156, 164–7, 170, 179, 187, *205*, 216
patents, 84–5, 91–8, 163–88, 212–13, 216
Patent and Trademark Office, USA, 188n, 213
Pavitt, K., 123n, 188n
Perelman, C., 66n
peripheral issues/keywords, *114*, 116, 120, 125, 166, 171, 193, 199, 201, 215, 226
Pinch, T. J., 20
pinocytosis, 72–6, 78–9, 83, 105
point of passage, 86–7
politics, 4, 10, 14–15, 90–1, 222–3
polymers, 143–8, 151, 153–4, 161–2, 191
Popper, K. R., 103
power, social, 4, 50, 221–2, 227
  of findings, 46
  of the laboratory, 11, 27, 35
  of science, 20, 98, 222–3
  of texts, 67, 223
Price, D. de S., 123n, 145
prince, 5, 98, 222–4
problematisation, *xvii*, 26–7, 75, *78*, 81, 87, 89, 108–10, 116–17, 122, 141n, 144–5, 160, 191
proximity coefficient/index, 112, *114–15*, 147–8, 153, 160, 192–3
proximity map, 114–16, 120–2, 133–40, 148–9, 156, 174–7, 179, 186–7, 215–16
punctualise, 73

qualitative, 11, 13, 103–4, 107–8, 122, 221, 223, 225–6
quantitative, 11, 81, 103–4, 107–8, 221, 223–6

rank, 114, *194*, 197
rats, 9, 11, 35–9, 75, 78, 105, 201
reader, xvii, 51, 57–8, 62, 64, 67–8, 71, 73–7, 80, 84–5, 92, 108, 223, 227
reductionism, 7–8, 15, 77, 104
Renault, 22–7, 30–2, 104–5
repertoire, 109, 119
research proposals, *see* funding applications
resources, 9, 11, 39–40, 46, 84, 94, 97–8, 153, 161
rhetoric, 51, 64, 92, 97
Rip, A., 98n, 99n, 103, 111, 193
role, xvi, 22, 26, 70, 105
Rothman, H., 211, 216
Rousseau, J-J., 6

scenario, 26, 52, 89
Schaefer, W., 103
Schmookler, J., 123n, 188n
Schumpeter, J. A., 162
science, 3, 6, 8, 91, 97–8, 104–5, 222–3
science indicators, 123n, 216
science policy, 97, 116, 142, 144, 160, 162, 212, 217, 222, 224
science policy-maker, 3, 12, 224, 226
scientific article, 10, 51, 65–6, 69, 80, 84–5, 90, 93, 106, 108, 144, 190, 198, 205, *see also* scientific paper
  journal, 13, 47, 68, 70–2, 207, 212–13
  method, 19
  paper, 11, 13, 43, 47, 67–8, 71, 73, 76, 81
scientists, 9, 20, 29, 35, 39, 42, 70, 80, 91, 104, 111, 226–7
scientometric analysis, 97
scientometrics, 13, 15n, 103
semiotics, 51, 66–7
Shapin, S., 103
signal words, 110, 225, *see further* key-words
simplification, xvi, *28–32*, 34, 40, 42, 46, 110, 115, 117, 140
skeleton graph, 194–7, 200, 202, 205
Small, H., 15n, 123n
Smith, M. R., 229
society, 4–5, 8, 14, 20, 33, 222, 227–9
sociologists, 20, 22, 66, 227
spokesman, xvii, *24–6*, 89, 105, 107, 228–9
statistical coefficient, 193, 197
strategic locus, *4*, 35, 222–3, 229
strategy, xvii, 11, 15n, 26, 92, 97, 99n, 107–9, 144, 149, 151, 161–2, 164, 184–6, 216, 223–5, 228
Strum, S., 228

technical object, 23, 28
technologist, 20–1, 104
technology, 6, 42, 104–5, 187, 213, 223, 229
texts, 10, 12, 14, *49–50*, 62–3, 67, 70, 73, 77, 79, 82n, 84–5, 90–2, 97–8, 105–9, 111, 176, 197, 205, 213, 223–6, 229–30
Thackray, A., 103
threshold, 113, 133, 148–9, 156, 164–5, 167–9, 171–3, 189, 192–3, *195–7*, 214, 228
Toulmin, S. E., 103
Touraine, A., 4
translate, xvii, 24, 28, 39, 49, 68, 70, 74, 86, 105, 156, 198, 228
translation, *xvii*, *24–8*, 46, 67, 69, 71, 74–7, 80, 82n, 88, 106–9, 111–12, 119, 122, 166, 170, 191–2, 211, 223, 225, 228–9, 230n
translation centre, *xvii*, 27, 35, 228–9
  chain, *see* chain of translation
  strategy, 156–7, 159–61, 228
transportable, 49, 229
Turner, W. A., 105, 123n, 143, 192

university, 143–4, 149, 152–3, 155, 157, 159–61

Van den Belt, H., 99n
Van den Daele, W., 103
VEL (véhicule électrique), *see* electric vehicle

Weber, M., 4, 7, 229
Weingart, P., 103, 142
Williams, R., 9, 52, 66n, 69, 81n
Woolgar, S., 9, 20, 66n, 82n, 103, 123n
words, 12–13, 68, 72, 74, 81, 91, 106, 108, 123n, 133, 136–7, 162n, 185
world of science, 9–10, 11–13, 50, 223

Yearley, S., 66n
Young, B., 7
Yoxen, E., 20

zoom, 168–73, 187, *193*, 215–16, 227